Springer Mineralogy

The Springer Mineralogy series publishes a broad portfolio of scientific books on Mineralogy, Petrology, Crystallography and Gemology. Springer Mineralogy is aiming at researchers, students, and everyone interested in mineralogy and petrology. The series includes peer-reviewed monographs, edited volumes, textbooks, and conference proceedings

More information about this series at http://www.springer.com/series/13488

Olga Alexandrovna Doynikova

Uranous Mineralogy of Hypergene Reduction Region

Using electron microscopy data

Springer

Olga Alexandrovna Doynikova
IGEM RAS
Moscow, Russia

ISSN 2366-1585 ISSN 2366-1593 (electronic)
Springer Mineralogy
ISBN 978-3-030-67182-2 ISBN 978-3-030-67183-9 (eBook)
https://doi.org/10.1007/978-3-030-67183-9

This Springer imprint is published by the registered company Springer Nature Switzerland AG.
The registered company address is: Gewerbestrasse 11, 6330 Cham, Switzerland

In memory of Larisa Nikolaevna Belova, an outstanding scientist-mineralogist, creator of the classification of oxidation zones of uranium deposits, researcher-romantic, passionate about scientific knowledge.

Annotation

This book comprehensively reviews the earlier underexplored uranium ore mineralization—uranium blacks/soots and its minerals of tetravalent uranium. Regularities of hypergene uranium mineralization, combining all types of ores, are generalized and schematized. The mineralization of oxidized U^{6+} (uranyl) and reduced U^{4+} (uranous) forms is considered a manifestation of a single redox hypergenesis zone. Examples of different genetic types of uranium deposits, where U-blacks are the mineralization of industrial interest, are given. The mineral composition of black (sooty) ores is detailed by using crystallochemical methods such as analytical electron microscopy (AEM).

New data on minerals of tetravalent uranium are obtained; for the first time all currently available information on ningyoite and other uranous phosphates has been systematized and generalized. An overview of all now known deposits of ningyoite ores is given. Uranium ores from paleochannel sandstone-type deposits of the Vitim ore region, Russia, with a previously little-known new type of uranium ore—phosphatic black, are described. The formation of uranous minerals in sedimentary strata was considered on the basis of the concept of biocos systems in light of geochemical barrier doctrine. The conclusion about the biogenic formation of black U-ores was reached when the phosphorus source for ningyoite was found.

The AEM method is the most effective tool for studying loose materials, and its data are the basis of the conclusions, so methodic section is given. Here the practice of electron diffraction (SAED) study of fine, poorly crystallized (including metamict) minerals is described, given by concrete examples of U-minerals. Such methodical consideration will be useful for different profile scientists to study highly dispersed materials. The generalization of uranium sooty ores study results is of interest to specialists in geology, mineralogy, and uranium crystallochemistry.

The research was carried out under budget financing for the topic of state scientific task for IGEM RAS.

Introduction

The nuclear industry has an important place in the development of power engineering worldwide. The most promising sources of raw materials for nuclear power are the uranium deposits suitable for exploitation by underground in situ leaching (ISL) method. Today this method is predominantly used to extract sandstone-type deposits. Such infiltration deposits (roll-type, basal channel, tabular, etc.), where uranium blacks (sooty) compose the cementing material of sandstones, are widely known.

Uranium black is a finely dispersed product of U^{4+} mineralization, formed in reductive conditions of weathering redox processes. In the supergene region, uranium blacks are present practically in all types of uranium deposits. Black mineralization is widespread in secondary enrichment zones of hydrothermal (endogenic) deposits and in sandstone-type deposits. In the Earth's sedimentary cover uranium blacks are of great interest in the industry because of their economic appeal. The wide prevalence and manifestation of the ores of various genetic types have undoubtedly generated scientific and practical interest in the mineral composition of these uranium ore formations, which are difficult to diagnose by traditional methods of mineralogy.

The term "infiltration" deposits is used by the author as an analogue of the widely used concept "exogenous-epigenetic deposits,", because this is a more detailed characteristic of hydrogen deposits, which clarifies their genesis.

Black mineralization is often the main and even the only ore component of industrial uranium deposits. The monograph contains examples of sandstone uranium deposits of different genetic types, on the basis of which black mineralization is developed in volumes of industrial interest. These deposits are Taboshar (Middle Asia, Karamazar ore region), Kosachinoye (Grachevskoye ore field, Northern Kazakhstan), Chu-Sarysu infiltration-type deposits (near Tien-Shan ore mega province), and basal channel (paleo-streams) deposits with a new type of phosphate black mineralization (Japan, Bulgaria, and Russia). The task of developing the uranium resource base due to new deposits of this type, with black ore mineralization, remains relevant at present.

This work is devoted to the mineralogical study of dispersed uranium minerals in the reductive hypergenesis region. In the study and development of hydrogene deposits by the ISL method, the diagnostics of uranium mineral forms refers to problematic issues. The ores of such deposits are usually represented by friable, loose material, and the ore substance itself is highly dispersed. Therefore, among the methods for studying this type of ore, electron microscopy (EM) occupies a special place, allowing us to diagnose uranium minerals at the level of micron-sized formations. To date, such a detailed mineralogical analysis of uranium black's composition has not been conducted abroad. The involvement of local methods in the study of dispersed uranium ore formations, in essence, brings to a new level the mineralogical study of uranium ores, allowing us to solve problems not only applied but also genetic.

The main method in our study of the mineralogy of uranium blacks is analytical transmission electron microscopy (ATEM). This work is largely based on a crystallochemical characteristic that is obtained directly during ATEM research, namely, the combination of diffraction data on a micro mineral structure and the results of its composition analysis. As EM did not become the standard method of mineralogy yet, the short review of EM methods with the description of their opportunities when studying dispersed mineral substance is provided in the book. The basic investigation method is selected areal electron diffraction (SAED) (in Russian—microdiffraction). Unlike X-ray methods, the technique for interpreting electron diffraction data is not unified, so its use in each specific case can be considered as a kind of methodological guidance in researches done by using microdiffraction. In this regard, the methodical section, which describes the techniques for deciphering microdiffraction SAED patterns, occupies a significant portion of the book; it then proceeds to sections devoted to the main results of the work. Considering that it is necessary and important for mineralogy to illuminate the experience of EM studies of dispersed mineral substances, the author found it possible to give so much attention to the main methods of research.

Analytical scanning electron microscopy (ASEM) has also been used in research work on uranium disperse mineralization study. The combination of different EM methods has contributed to development knowledge about regularities of uranium mineral formation in the hypergenesis region. The characteristic, for this environment, friability of samples and dispersity (powder-like) of studied substance make electron microscopy an indispensable method for mineralogical research.

This book sets out results of loose ores study from the reductive hypergenesis region, which represent well-known but still poorly studied ore formations—uranium blacks. Earlier, the predominant opinion was that uranium minerals here was exclusively present in the form of oxides, and the presence of the silicate form (coffinite) has been considered rather a rare feature of uranium blacks. Only in recent decades, coffinite, along with pitchblende, is mentioned as common uranous ore component. Only local EM methods to study the mineral composition of uranium blacks made it possible to detect a new phosphate uranous mineral form.

Oxide uranium mineral form, usually predominant in black's composition, has been carefully studied earlier, and uranium silicate, coffinite, has been also widely studied. As for uranous phosphates, which are part of uranium black's composition, their mineralogy and crystallochemistry were previously in fact unknown. Previously unknown minerals U^{4+} were also discovered with the help of ATEM.

The characteristic feature of all the minerals considered here is the dispersity of their natural formations. The revealed crystallochemical features of finely dispersed ore minerals, which form uranium blacks (oxides, phosphates, and silicates of tetravalent uranium), allow us to understand geochemical conditions of their formation; it helps us to resolve mineralogy problems in a new way. The study of this dispersed uranium mineralization both in secondary enrichment zone and in various types of infiltration deposits allows us to speak about the similarity of conditions and regularities for black ore formation. Data on the crystallochemistry of uranium minerals reflect the features of mineral formation and allow one to determine the laws of migration and the deposition of ore components (in the form of minerals) within the hypergenesis region.

The established crystallochemical features of uranous minerals (in black's composition) are indicators of geochemical medium and allow the qualitative determination of the ore process conditions. The detailed nature ore material study makes it possible to more accurately determine the behavior of uranium, both in natural and in technological processes. Systematization of crystal chemical studies results of dispersed uranium minerals not only expands mineralogical data on uranium ores (of industrial interest), but also leads to deepening knowledge on processes of this ore formation in the hypergenesis region.

This book summarizes the author's results of research carried out since 1976 in Exogenous ore formation Division in Radiogeology and Radiogeoecology Laboratory, IGEM RAS, named after D.I. Shcherbakov. All the research was carried out thanks to budget financing for the topics of research of the state tasks of IGEM RAS during all these years.

The work was carried out under Dr. L.N. Belova on samples and materials mainly from her personal collection. Material composition study of exogenous uranium ores has been carried out using analytical TEM and SEM methods. These experimental investigations were performed under Dr. A.I. Gorshkov in the laboratory of electron microscopy IGEM RAS (now N.V. Belov laboratory of crystal chemistry of minerals). The author is deeply grateful to Prof. A.I. Gorshkov for teaching the method of electron microdiffraction and for participating in research. The EM method possibilities in uranium mineral study are really manifested only due to high professionalism of the unique uranium mineralogist L.N. Belova due to her precise selection of key "bench marks" as research objects. She was well-known Russian researcher of uranium deposit oxidation zones. A tribute to the memory of her is author's attempt to geologically understand and interpret obtained results; this is based on L.N. Belova's ideas, in particular, the taxonomy of uranium deposit oxidation zones and systematization of hypergene uranium ore formations. The author is grateful to the staff of D.I. Shcherbakov Laboratory: Corresponding Member of RAS Velichkin V.I. and PhDs Omelianenko B.I., Borisenko E.N., and Timofeev A.V. for discussing study results, over critical remarks, and valuable consultations on geology and mineralogy uranium. I thank all the staff of Laboratory Electron Microscopy for their constant support, especially PhD Samotoin N.D. for his valuable advice and A.V. Sivtsov for his invaluable assistance in carrying out experimental research.

Contents

Chapter 1
Practices of Dispersed Uranium Minerals Study by ATEM Methods

1.1 Transmission Electron Microscopy Possibilities in Minerals Study

The main method for minerals identification, on which all the research results presented in this book are based, is the method of Selected Area Electron Diffraction (SAED), carried out in transmission electron microscope (TEM). Exactly this method is the most effective in the study of dispersed substance crystallochemical characteristics and optimal for obtaining complex of crystallochemical data sufficient for diagnostics, in combination with elemental analysis of the same substance directly into the TEM.

TEM is used in mineralogy to solve problems, associated first of all with reliable phase diagnostics of dispersed micro- and nano-dimensional mineral formations, which poorly crystallized, or complex polymineral formations (including mixed, disordered, metamict). TEM methods effectively solve the problems of phase mineral transformation (replacement, weathering); problems of growth and dissolution of minerals, associated with detailed study of the crystal surface; questions of mineral heterogeneity at the micro level (intergrowth, twinning and epitaxy), etc. TEM is one of the main research methods in materials science. Various TEM methods observe and investigate defects in the crystal structure, and study the structure of thin films.

The wide use of TEM in various fields of science and applied fields: solid-state physics, biomedicine, chemistry, mineralogy, materials science, nanotechnology, etc.—stimulates the development of a variety of specific preparation techniques adapted to the tasks and objects of research. In mineralogy there are two main preparation methods to study by TEM—replicas and suspensions (Gritsaenko et al. 1969). Until to the equipping TEM by analytical techniques, widely used replicas (coal or coal-metal)—relief volumetric films, surface imprints, which make it possible to examine the smallest micron details and the relief features of the sample under study with resolution order of 10 Å. The study of morphology is combined

O. A. Doynikova, *Uranous Mineralogy of Hypergene Reduction Region*, Springer Mineralogy, https://doi.org/10.1007/978-3-030-67183-9_1

with diffraction study of substance was extracted on the same replica. When suspension preparation the study objects are thin, transparent to electron beam, micro- and nano- scale mineral particles. Suspensions are prepared by ultrasonic dispersion of mineral grains in distilled water; then drops of the suspension are applied to a special preparation grid covered with an electron-transparent film-substrate (collodium-coal, etc.) and dried. The development of new preparing sample methods for TEM research is most active in field of studying polymeric materials and moisture-containing objects (cryogenic methods development for ultrafast freezing).

For uranium minerals study by replica method, a preparation technique was developed based on contrast radiography—EMAR (electron microscopic auto-radiography), which allows to detect dispersed radioactive phases and even their non-mineral form, but with low radiographic resolution (20–50 μm). This method detailed description is given in monograph (Sidorenko et al. 1986), where minerals preparation experience (ultramicrotome, ion thinning out and chemical etching) is generalized and rational EM complex of uranium minerals and ores studies was evaluated. Among all mineralogical study methods of radioactive ores, EM is considered as one of the most promising.

Abroad, HRTM (High Resolution Transmission Microscopy) is increasingly used for mineralogical research. Resolution 1–3 Å (which is comparable with atoms size) allows to obtain a direct image of the crystal structure with all its disturbances (lattice disorder, defects, dislocations, vacancies, atoms displacement from ideal position, etc.). For this purpose significant role is played by computer simulation image of crystal structure. A good example of HRTM importance in solving genetic Mineralogy problems is the study of phyllosilicates in hydrothermally altered oceanic basalts (Shau and Peacor 1992). The authors concluded that mixed-layered metastable minerals appear only when formation of relatively stable individual phases is suppressed. This indicates rapid crystallization or minerals formation at low temperatures in absence of significant fluid quantities. There is emphasizes the necessity of HRTM image applying (along with the X-ray diffraction!) for unambiguous diagnostics of disordered minerals.

There are many theoretical publications and practical manuals that explain the physical basis of electron diffraction (ED) and diffraction methods in TEM (Vainshtein 1956, 1964; Hirsch et al. 1968; Andrews et al. 1971; Drits 1981; Dorset 1995, etc). The structural analysis of nanomaterials using ED to obtain three-dimensional information and high data completeness is now widely known (Kolb et al. 2007). On last decades electron diffraction methods in TEM are in progress owing to newest instrument equipment; process of electronograms shooting is currently sufficiently automated, how it has already been long done in X-ray diffractometry. Precession electron diffraction (PED) is a specialized method to collect electron diffraction patterns here. By rotating (precessing) a tilted incident electron beam around the central axis of the microscope, a PED pattern is formed that is suitable to determine the crystal structure of the sample (as help to direct methods). The situation has changed rapidly in the last decade with the advent of electron diffraction tomography (EDT); this method collection of a series of randomly oriented ED patterns at a fixed angular interval (Kolb et al. 2007). Comprehensive open-access crystallographic databases are offered for support of the automated crystallite phase identification process (Moeck et al. 2011).

Modern TEM allows us to carry out investigations as in usual (static) electron beam mode (transmission light) to obtain light—and dark-field images and diffraction patterns, and in scanning mode (transmission or reflection), which allows conduct element analysis. Computer equipment and software for image processing allow obtaining particle size and shape distribution, element distribution mapping, etc. The main advantage of analytical TEM (ATEM) is the ability to obtain data sets both in composition (spectra) so in structure (diffraction patterns) on the test substance, which ensures its reliable diagnosis; it is SAED method that the author has used.

1.1.1 SAED Method in Dispersed Minerals Study

In practice of mineralogical research, electron diffraction is widely used for phase analysis of dispersed mineral mixtures and determination of minerals structural characteristics. Analysis of electron-diffraction reflexes intensities is used to reveal crystal-chemical features of minerals structure. This is the main method of crystal structure studying for the powdery, weakly or poorly crystallized and metamict objects.

Electron diffraction as a method for solids structure studying is based on B. K. Vainshtein's works (1956, 1964). The detailed explanation of the physical foundations of electron diffraction, as well as the principles of geometric analysis of diffraction patterns of single crystals, is given in a number of articles and monographs, the basic of them are (Vainshtein et al. 1979; Pinsker and Goodman 1981). In development of structural analysis, based on electron diffraction, a significant contribution was made J.M. Cowley (1967), by theoretical and experimental work. The electron diffraction theory, principles and theory of electron diffraction patterns calculating are detail discussed in a number of monographs (Hirsch et al. 1968; Andrews et al. 1971; Schimmel 1972; Utevsky 1973; Vainshtein 1979; Drits 1981, 1987, and others). L. M. Utevsky (1973) described in detail the method of calculation and indexing of electron diffraction patterns, on examples of metal phases, which are characterized by symmetry, not lower than orthorhombic. In the case of mineral objects interpretation methods is much more complicated, since the minerals structures represent almost all classes of symmetry.

Historically, using of electron diffraction in mineralogy began from *method of electronography*. Diffraction patterns in electronograph obtained not from individual mineral particle, but from a larger number of particles with the same orientation, that creating so-called "textured" object of study. Usually, it is oriented preparation from layered, lamellar minerals. The most informative are the electronograms of oblique (skew) textures that contain three-dimensional set of hkl reflections that are distributed (in two dimensions) over series of hk-ellipses. Such electron diffraction patterns from oblique textures are still used to accurately diagnose and structural polytypes analyze of layered minerals.

In Russia the first generalization of minerals studying practice by electron diffraction was carried out in work (Zvyagin 1964), where electron diffraction data on the structural crystallography of layered silicates were systematized. Electron diffraction methods in application to mineral objects were discussed in one of the first monographs devoted to the EM in Mineralogy (Gritsaenko et al. 1969). Illustrative examples of SAED application to detailed study of crystallochemistry for dispersed mineral formations are numerous publications of A.I. Gorshkov with coauthors on Fe-Mn oxides-hydroxides study; that was generalized in the monograph (Chukhrov et al. 1989). The physical features of electron diffraction make it possible to carry out structural analysis of highly dispersed (powder-like) and poorly crystallized mineral objects, which was demonstrated in V.A. Dritz's works on study of mixed layered phyllosilicates (Drits 1981; Drits et al. 1993) and phyllomanganates (Drits et al. 1997, 2002).

Since beginning 1990s, a new electron diffraction direction is come into service: Electron Diffraction Structure Analysis (EDSA); the term was introduced by B.K. Vainstein to determine the method for deciphering crystal structures using electron diffraction, in contrast to X-ray or neutron diffraction. The article (Vainshtein et al. 1992) reviews the state of EDSA method and the main directions of its practical application for period from its first steps (1950) to 1991. The review includes theoretical background and description of the preparation methods; here structural studies results based exclusively on electronic data are summarized.

The first stage of structural electronography analysis is geometric analysis of electron diffraction point patterns, results in determination of space group symmetry and elementary cell parameters of mineral. The next, more complex deciphering stage is analysis of diffraction reflexes intensities, which will make it possible to reveal thinner crystallochemical features and to determine properly mineral structure. Crystal structure deciphering using intensity of diffraction reflections is the most difficult task of electronic diffraction, which, in addition to modern software, requires considerable time to obtain necessary amount of initial data. The intensity of electronic-diffraction reflections characterizes a structure of substance via distribution in it electrostatic potential (in x-ray diffraction, this corresponds to the function of electron density). Obtained experimentally quantitative intensities estimation is the basis for deciphering of mineral structure. EDSA is an exceptionally powerful method of researching structures by its technical capabilities.

Review of structural data on inorganic crystals reports that in recent years, only a few structures, i.e. about 0.6% of all newly deciphered ones, were determined either completely by electron diffraction data or with their help (Weirich 2003). Features and capabilities EDSA method discussed in the article (Avilov 2003a), where high assessment of potential for combining it with HRTM to structural's studies was given. It is in this combination ATEM (with diffraction capability) and HRTM, which allows crystallochemical study of micro—and nano—dimensional minerals, it seems future prospects of electron crystallography development.

At present, structural studies based on electron diffraction with calculation of electrostatic potential (ESP) distribution are few in Russia (Tsirelson et al. 1998; Zhukhlistov et al. 1997). The first results of quantitative ESP study are presented in review of recent works that develop precision electron diffractometry and methods for analyzing of ESP distribution (Avilov 2003b). Works on structural (including ab-initio) definition and structures refinements of inorganic compounds (i.e. minerals in the powder monofraction form) by electron diffraction methods are known (Weirich et al. 1998, 2000, 2002; Huang et al. 1996). These methods for processing electron-diffraction data are developed using modern programs.

Information we may receive. Through the EDP (similar to the way it is done by means X-rays) the structural characteristics of mineral are determined: parameters and unit cell type, point symmetry group and possible space group. It is possible to appreciate degree of crystallinity and order-disorder of structure. It is possible to study structural features of minerals: structure dislocations, packing defects, twinning, etc.

Further, the author in accessible form (in a volume necessary for understanding future mineralogical content) briefly considers the basic principles of analytical EM, gives a brief description of SAED method, and discusses comparative possibilities of electron and X-ray diffraction. The principles of EDP decoding and methodic techniques, which the author used to unit cell parameters determine, are shown on example of specific minerals.

For in-depth acquaintance with SAED physical fundamentals, general principles of decoding electron diffraction patterns, methods for constructing reference point electron diffraction patterns, etc. reader should refer to modern methodological manuals on solid state physics.

In this work SAED method underlies diagnostics of dispersed uranium minerals, which are represented both in the composition of disperse polymineral mixtures and in monomineral sedimentations extremely small sizes. Such mineral formations are difficult or inaccessible to X-ray diffraction methods of diagnostics. In combination with elemental composition analysis, SAED forms an effective complex of substance research crystallochemical methods, which is very effective for mineralogical studies. Further in the text following terms are used for SAED pictures naming as synonyms—electron diffraction pattern (EDP) or electron diffractogram (ED).

1.1.1.1 Comparative Possibilities of Electron Microdiffraction and X-Ray Diffraction in Dispersed Uranium Minerals Study

When studying the structural characteristics of dispersed minerals two diffraction methods are usually used: X-ray and electron (SAED). Samples of exogenous uranium ores, studied by the author, in most cases presented a powdered mineral mixture. The uranium mineralization itself is a fine-dispersed formation, which is difficult to diagnose with usual light optics, and their content in the sample is often too small for analysis by x-ray diffraction method. Therefore, the question about comparative possibilities of these two diffraction methods—X-ray and SAED—is the actual in study of crystallochemistry on complex fine-dispersed mineral formations.

The main difference between electron diffraction and X-ray diffraction lies in the physical nature difference in diffracting radiation. The wavelength for electrons with energy diapason of 30–200 eV (as in EM) is approximately 0.1–0.2 nm, which satisfies the diffraction condition for atomic structures, because the wavelength is equal to or less than interatomic distances. Advantages of electron diffraction (in comparison with X-ray radiation) are determined, first of all, by shorter wavelength, and by stronger (on 3–4 orders) interaction of electrons with matter (Vainshtein 1956; Drits 1981), which allows identification of crystals very small size. Electron diffraction investigation methods make it possible to obtain structural characteristics of X-ray-amorphous minerals, since electrons wavelength can be several orders smaller than that of X-rays. SAED efficiency for study X-ray-amorphous matter has been repeatedly confirmed in the author's practice, both in study metamict brunnerite and in study highly dispersed uranium oxide (as part of uranium pitchblende). SAED advantage is also possibility of obtaining single-crystal (point) diffraction patterns of microcrystal (up to 0.1 μm in size). SAED using allows to search individual mineral phases in a polymineral mixture and to do their diagnosis.

Unlike XRD, a feature of ED is the ability to observe SAED-patterns directly on electron microscope screen. We can correct diffraction image by manual adjustment or by tilt of diffracting object itself for pattern improving: namely, for achieving symmetry as in geometry of reflexes location and in distribution of their intensities. These manipulations serve as a kind of "adjustments" of diffraction image. It is these most symmetrical EDPs are fix on photo plates and are used for further calculations.

Electron diffraction is not as accurate as X-ray diffraction, but it has its advantages when increasing degree of substance dispersion. Practically it appears as follows: during diffraction studies of the same loose material by rays of different nature, the electron diffraction patterns reveal a higher crystallinity degree of the studied material than reveal X-ray ones. Namely: the X-rays patterns with clear ring reflexes (for example, uraninite) characterize a sample in which the predominant part of particles give (produce) a point ED pictures. Diffused X-ray rings on patterns correspond to near equal ratio in a sample of particles with ring and point SAED-pictures. In the suspension preparation from X-ray amorphous (uranium) mineral, as a rule, particles that give ring electron diffraction predominate; along with this, there are also particles with point SAED-patterns. As the limitative variant of crystallites dimension, one should consider electron-amorphous uranium oxides (in uranium blacks) that give wide diffraction halo or two extremely diffuse wide rings on the SAED-pictures. Diagnostics of such electron-amorphous formations (certainly—X-ray amorphous!) is possible only from elemental analysis of mineral particles composition.

In addition, in APEM—studies of multicomponent polymineral mixture, it is possible to choose particles of the certain phase in a preparat; moreover, there is a possibility of selecting particles that give the best EDP quality. It is this microscopic "localness" of crystallochemical studies that explains advantage of EM in mineral mixtures or dispersed objects study.

A possibility to choose particles, the most perfect structurally is extremely essential when studying metamict mineral objects. In this way X-ray amorphous metamict brannerite was studied: among many particles in the (suspension) preparation, those that were least changed and retained ability to diffract were selected. Thus, for the first time were obtained the structural characteristics of natural brannerite without pre-calcination, traditional for metamict minerals (Ivanova 1982; Ivanova et al. 1982).

Above mentioned clearly show a promising outlook of SAED using in mineralogy of finely dispersed, poorly crystallized and inaccessible for X-ray diffraction, uranium minerals, which are often objects of interest at uranium exogenic deposits.

1.1.1.2 General Principles and Techniques for Electronograms Interpretation

The principles of EDs interpretation are borrowed from x-ray diffraction method, with modification of scattering atoms function (electron radiation instead of x-ray). Reciprocal lattice is a conditional symbolic image of the mineral crystal structure, in which geometry of diffraction peaks corresponds to nodes location in real (direct) lattice of the mineral, and diffraction intensity of these nodes is determined by the location and by sort of the atoms in the real mineral structure. The basis of the decryption method is the concept of inverse "reciprocal lattice" of the crystal structure. The translations periods of direct and reverse lattice are characterized by mutually opposite values. Given the multiplicity of crystal structures, there is no universal method of decoding SAED-patterns, and therefore, each mineral requires an individual methodological approach. Concrete examples give a contribution the practice of methodological techniques for decoding electronograms.

By V.K. Vainshtein definition, "the electron diffraction pattern is a plane section of the reciprocal lattice that passes through its beginning of coordinates, on the scale Lλ" (Vainshtein 1956). The diffraction pattern is the Fourier-image of the object. In electronograms, the interplanar spacing d corresponding to the diffraction reflex is calculated by the formula $rd = L\lambda$, from which $d = L\lambda /r$, where r—is a distance measured from center of diffraction pattern to the reflex, measured on a photo plate; L—is a distance from sample to photo plate (effective length of a chamber); λ—is wavelength. The plane of the reciprocal lattice (a point diffraction picture, depicted on an electronograms) is usually denoted by a symbol (uvw) *. The values of u, v, w are determined by indices of crystallographic axis of [uvw] zone in direct lattice. This axis is perpendicular to electron diffraction pattern, i.e. is parallel to microscope optical axis. Each node of a reciprocal lattice is corresponds to a series of planes (hkl) in a direct lattice and it is denoted by the same index—h, k, l. Individual reflexes in the electronogram plane can be indexed if parameters of the crystal cell are known. All reflexes in a certain plane (uvw) must have indices hkl, corresponding to the equation of plane $hu + kv + lw = 0$. The index of the electronogram plane is determined by two reflections $(h_1k_1l_1)$ and $(h_2k_2l_2)$, which don't lie on the same straight line, by known method of matrix multiplication (Hirsch et al. 1968; Andrews et al. 1971).

At high symmetry of reflexes location (e.g., with orthogonal geometry of diffraction pattern) basic cross sections of reciprocal lattice (with simple indexes) are diagnosed relatively easily. Indication of reflexes and determination of unit cell parameters are carried out on the basis of extinction law, i.e. existence of systematic extinctions among the intensity maxima. Such extinctions exist throughout an entire inverse space for non-primitive lattices, regardless of the crystal symmetry. Additional extinctions occur when there are certain spatial symmetry elements. So, for example, planes of sliding reflection cause extinctions in corresponding planes of reciprocal lattice; and helical axes cause extinctions along isolated rows of reciprocal lattice. Data on space groups and connected with them extinctions are given in many special reference tables.

Depending on degree of studied material crystallinity, EDPs can be point-like (in the case of micro monocrystal) or annular (if diffracting microparticles are composed of polycrystalline aggregate of nanoparticles). Ring SAED-pattern is a flat cross-section of a system of diffraction rays formed by polycrystal. Point-like MD-picture represents a cross section plane of diffraction nodes in reciprocal space and indicates monocrystal nature of studied particle. Symmetrical, both in reflexes location geometry and in their intensity, SAED-patterns demonstrate undistorted plane of reciprocal lattice of micro monocrystal.

The intensity of diffraction rays $\boldsymbol{I}$ (manifest in blackening diffraction spots degree) serve as visual expression of experimental scattering ability of crystalline object. The theoretical values of hkl reflections intensities are calculated as quantities proportional to structural amplitude F(hkl) with allowance for corrections (on mosaicity, blockiness, etc.): for kinematic scattering $\boldsymbol{I} = F^2$; for dynamic scattering $\boldsymbol{I} \sim F$. The structural amplitude, or structural factor F, is the sum of the scattering amplitudes by all the atoms of a unit cell. For the known structure, or its model, calculates theoretical values of the intensities and the resulting experimental intensities are compared to these theoretically calculated.

In determining of minerals crystal structure parameters by SAED method sets of point EDPs obtained from one microcrystal was used. These sets were received in microcrystal rotation by goniometer and, thus, all EDs were connected by a common axis of rotation. The theoretical foundations of such method (of interrelated EDPs) have long been developed (Vainshtein 1956, 1964; Utevsky 1973; Zvyagin 1964), but their use for studying previously unknown crystal structures (i.e. minerals) has not received wide propagation. In other countries this approach is used in a limited way and as a rule it used only for objects with already known structure, and which are accessible for X-ray research.

When shooting series of point EDPs (EDs) with common rotation axis usually one of the crystallographic coordinate axes selected. Rotating the crystal in SAED mode, we see changing the diffraction pattern on EM screen. In other words, we observe SAED-patterns appearing at sequential (gradual) intersection of reciprocal lattice nodes by plane of diffraction image; i.e. we observe different cross sections of reciprocal lattice nodes. Observing, we select (and fix for further calculations) those

SAED-patterns on which there are point reflexes closest to EDP center. The maximum diapason of observation—120° (goniometer was provide tilt ±60°). By assemblage of such interrelated point diffraction patterns (namely, by the reflections nearest to the center) one can unambiguously determine the type of unit cell. Next it is indexation of reflexes and planes.

To determine the space group, i.e. revealing a complete set of systematic extinctions, a series of EDs (at least three), connected by the common rotation axis, is needed. It should be specially emphasized that such method allows one to obtain *monocrystal* (point) diffraction characteristics from submicron and more size particles.

1.1.1.3 Methodical Techniques for Decoding SAED-Patterns

Calculations of structural characteristics (unit cell parameters, lattice symmetry) were performed in cases when the microcrystals were tilted around several different crystallographic axes (for ningyoite, urphoite) and around one axis (for brunnerite, carnotite, uraninite, and coffinite). In order to reveal spatial location of reflexes, in all cases, sets of EDs are used, which was obtained by tilting monocrystal particles around main axes. These crystallographic axes were revealed by the SAED-patterns. As a rule, when shooting a series of point EDs, the author selects one of the coordinate crystallographic axes as the common axis of rotation.

Received SAED-patterns were considered as result of rotation of some (certain) cross section of reciprocal lattice (chosen as initial one) around tilt axis, which passing through the nearest to center reflexes ($h_1k_1l_1$). During imagined turn is made, other reflexes indexes ($h_2k_2l_2$) are geometrically determined. The indexes of electron diffraction plane (SAED-pattern) are determined by well-known method of matrix multiplication for two reflexes ($h_1k_1l_1$) and ($h_2k_2l_2$), not lying on the same ray. The author has performed graphical constructions that created the coordinate cross-sections of the studied crystals on the base of series of electronograms.

To construct the schemes, spatial reflexes were used whose vectors are perpendicular to the rotation axis. Such method of deciphering made it possible to determine following characteristics of studied uranium minerals: extinction laws, indexes of reflexes and reciprocal lattice planes of a crystal, the unit cell parameters, point group of symmetry, and possible space group. This principle of interpretation for point SAED-patterns is the basis for all microdiffraction studies presented below. When decrypting EDPs of minerals which syngony is orthorhombic or monocline, both experimental and literature data were used to construct theoretical cross-sections of the reciprocal lattice.

- *Decoding SAED patterns of minerals with cubic symmetry.*

To decode EDPs for minerals with known (or supposed) cubic symmetry of crystal lattice (uraninite, pyrite), traditional methods known in X-ray diffraction were used.

Interpretation of ring EDPs for cubic syngony is very simple. Analysis of reflections complete set on diffractogram reveals extinctions and allows one to reliably establish one of the three types of Bravais lattices, permissible for cubic system: primitive (P), volume-centered (I), face-centered (F). The set of diffraction rings reflects the regular sequence of permitted reflexes. The set of diffraction rings reflects regular sequence of permitted reflexes (Bokiy and Poraj-Koshits 1964). This set allows one to unambiguously determine parameter of cubic unit cell by formula $\boldsymbol{a} = d_{hkl}\sqrt{(h^2 + k^2 + l^2)}$, which expresses the interrelation of parameter, interplanar distance d_{hkl} and the corresponding indices hkl. For known value of parameter $\boldsymbol{a}$, reflexes indices are determined from the same relation $\sqrt{(h^2 + k^2 + l^2)} = \boldsymbol{a}/d_{hkl}$.

Despite the simplicity, however, in the diagnosis of various minerals with close values of the parameter $\boldsymbol{a}$, methodic difficulties are possible, due to accuracy limits of parameter determination by SAED method ($\Delta\boldsymbol{a} = \pm 0.03 \div 0.05$ Å). This problem arises when there is no elemental composition control of particles (i.e., there is no spectrometer in TEM equipage). For example, at SAED diagnosis of uraninite in a disperse mixture with pyrite (association with pyrite is very characteristic for pitchblende ores). Microscale complexity consist in fact that both minerals (in highly dispersed form) are characterized by ring SAED-patterns, both belong to cubic syngony and have close parameter values $\boldsymbol{a} \approx 5.4$ Å. The accuracy of SAED-patterns calculation does not allow determining existing difference in parameters. Therefore, pyrite (sp. gr. Pa3) and uraninite (sp. gr. Fm3m) give practically the same values of interplane distances d/n for two first reflections, most clearly manifested on ring ED-patterns. The only one difference is that pyrite ED—ring-pattern contains additional reflexes compared to uraninite ED-ring-pattern; these additional reflexes characterize corresponding primitive spatial group (without extinctions). When working with circular electronograms, it should be remembered that dispersity of studied minerals is manifested in reflexes expansion and weakening. It is affects primarily reflections of distant orders and low intensity reflexes. In practice of our many years research, it wasn't found uraninite with different type unit cell than Fm3m (despite such reports in literature).

Point (monocrystal) SAED-patterns (EDPs)—correspond to planes that cutting diffraction knots of lattice in reciprocal space. The calculations were made on assumption that closest to center symmetric reflexes belong to one of three possible simplest sections in reciprocal lattice of cubic system. Thus, reflections closest to center have been chosen with supposed indexes ($h_1k_1l_1$). Next, this point-like EDP was considered as a result of rotation of some initial (chosen as the original) plane of reciprocal lattice around the axis, that goes through nearest to center reflexes. With imaginary rotation, indices of other (not lying on this axis) reflexes ($h_2k_2l_2$) are geometrically determined. Thus, EDP has been designate with two reflexes ($h_1k_1l_1$) and ($h_2k_2l_2$), which are not lying on one ray. The reflexes indexes and correctness of indexing are checking according to known (or assumed) parameter a of cubic unit cell, proceed from the above formula $a = d_{hkl} \cdot \sqrt{(h^2 + k^2 + l^2)}$. Using these decoding techniques, number of electron diffraction patterns that represent various planes of inverse (reciprocal) lattices of pyrite and uraninite have been calculated by author.

The long-term experience of our studies uranium blacks confirms crystallochemical affiliation of natural uranium oxides to single mineral form—uraninite, as noted earlier (Kerr 1951; Sidorenko 1978, etc.). The results convincingly show that division of simple uranium oxides for uraninite, uranium pitchblende and uranium black, as used in mineralogy, reflects only their external, morphological character, diffractively all of them represent one mineral with cubic face-centered elementary cell Fm3m, i.e. correspond to structural characteristics of uraninite. From structure point of view, when it comes for diffraction, one should name all these varieties—uraninite, noting only different degree of structure perfection (crystallization degree).

- *Decoding EDPs of minerals with hexagonal symmetry.*

An example of determining structural characteristics of hexagonal symmetry crystal is decoding of EDP lots (series) obtained with rotation of *ningyoite* micro crystals around different coordinate axes (Belova et al. 1985).

Initially, in studying our first finds of ningyoite, it was found that orthorhombic mineral structure of the mineral is not primitive, as determined by Japanese researchers (Muto et al. 1959), but it is C-base-centered lattice (Belova et al. 1978a, b). The unit cell parameters calculated from series EDPs, within the method accuracy, did not differ from those established earlier by the discoverers ($\boldsymbol{a} = 6.78 \pm 0.03$, $\boldsymbol{b} = 12.10 \pm 0.05$, $\boldsymbol{c} = 6.38 \pm 0.03$ Å). Later, on the basis of detailed SAED studies with rotation of microcrystals, the unit cell parameters of ningyoite were refined by us. As result, it was proved that the true symmetry of ningyoite structure is hexagonal (Belova et al. 1985), and initially selected C-centered orthorhombic cell (with orthogonal crystallographic axes) is actually hexagonal one, "orthohexagonal" in terminology of G.B. Bokiy (1971).

Refinement of the ningyoite parameters was carried out on EDPs series obtained by rotating microcrystal around two mutually perpendicular crystallographic axes (c^* and b^*). These rotation axes were chosen respectively along and across the crystals elongation $\boldsymbol{c}$ axis. The angles between SAED-patterns (planes of reciprocal lattice) that are fixed upon crystal rotation are determined experimentally. Clearly expressed microcrystals elongation along the c axis contributed to successful experiment realization. Experimental angles values strictly correspond to theoretical value between analogous planes of reciprocal lattice (Fig. 1.1 and 1.2).

According to SAED study, parameters of orthorhombic C-centered (orthohexagonal) unit cell were determined: $\boldsymbol{a} = 6.86$; $\boldsymbol{b} = 11.88$; $\boldsymbol{c} = 6.38$ Å (±0,03). The refinement by method of least squares was carried out for 38 independent reflections of 0kl and h0l type; their interplanar distances have been measured on point SAED-pictures, which strictly correspond to basal cross sections of ningyoite. Parameters of hexagonal cell of ningyoite are: $\boldsymbol{a} = \boldsymbol{b} = 6.86$; $\boldsymbol{c} = 6.38$ Å (±0.03). To improve the accuracy of measuring interplanar distances, point-SAED pictures with "internal" standard (Au sprayed onto preparation) was used (Fig. 1.3). Analysis of reflexes intensities played important role in proof of ningyoite hexagonal symmetry. Among EDPs obtained by rotating crystals around elongation axis ($\boldsymbol{c}$ *), identical EDPs are sequentially fixed every 60° (Fig. 1.4).

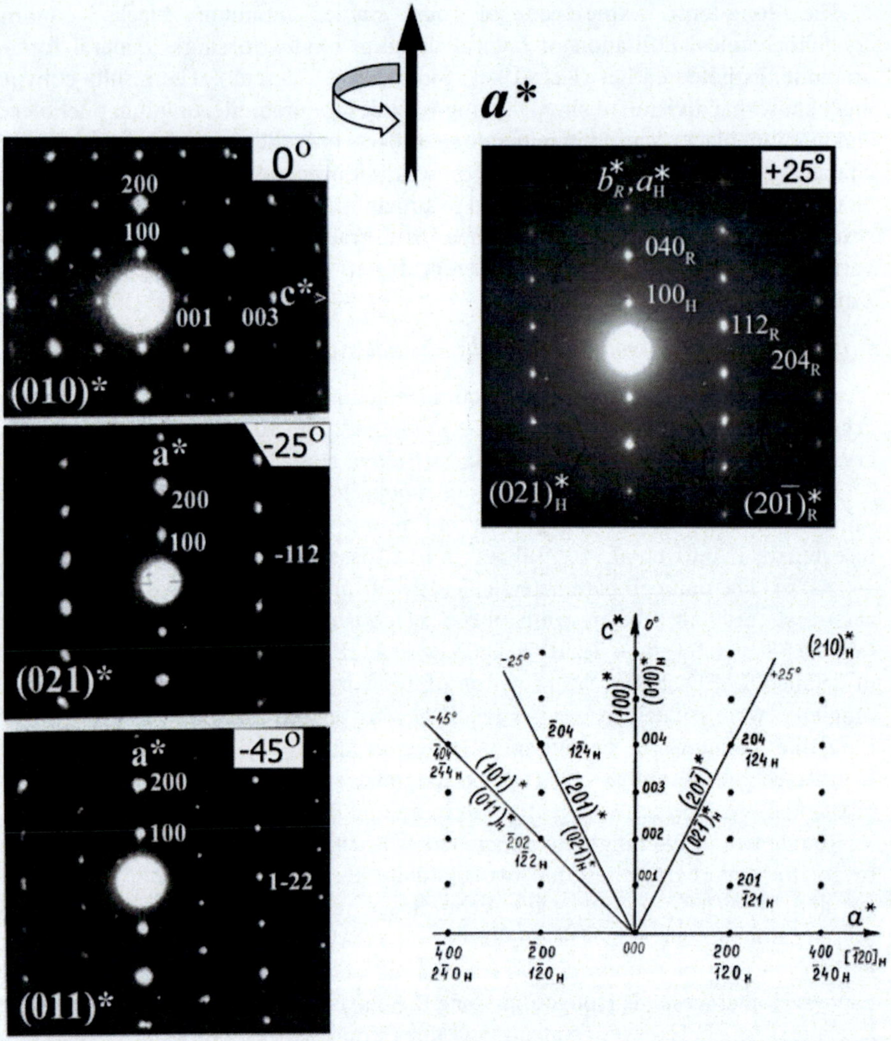

Fig. 1.1 Ningyoite. Series of electron diffraction patterns (EDPs) obtained with the tilt of microcrystal around ***a**** axis and their decoding scheme: graphical construction of plane, perpendicular to rotation axis. (On right—scheme and MDP—initial indexing carried out in orthorhombic syngony; with index H—indexing in hexagonal one)

These planes of reciprocal lattice are identical both in geometry of reflexes location and in intensity of reflexes, with the same value of d/n, which indicates true hexagonal symmetry of ningyoite.

The absence of reflexes extinction in hexagonal symmetry characterizes 8 spatial groups: three groups of the diffraction class C_{6h} and five—of diffraction class D_{6h}. One of ED-pattern obtained by tilting around ***c*** * axis (right on Fig. 1.3) demonstrates the absence of symmetry plane **m** perpendicular to sixth-order axis, which

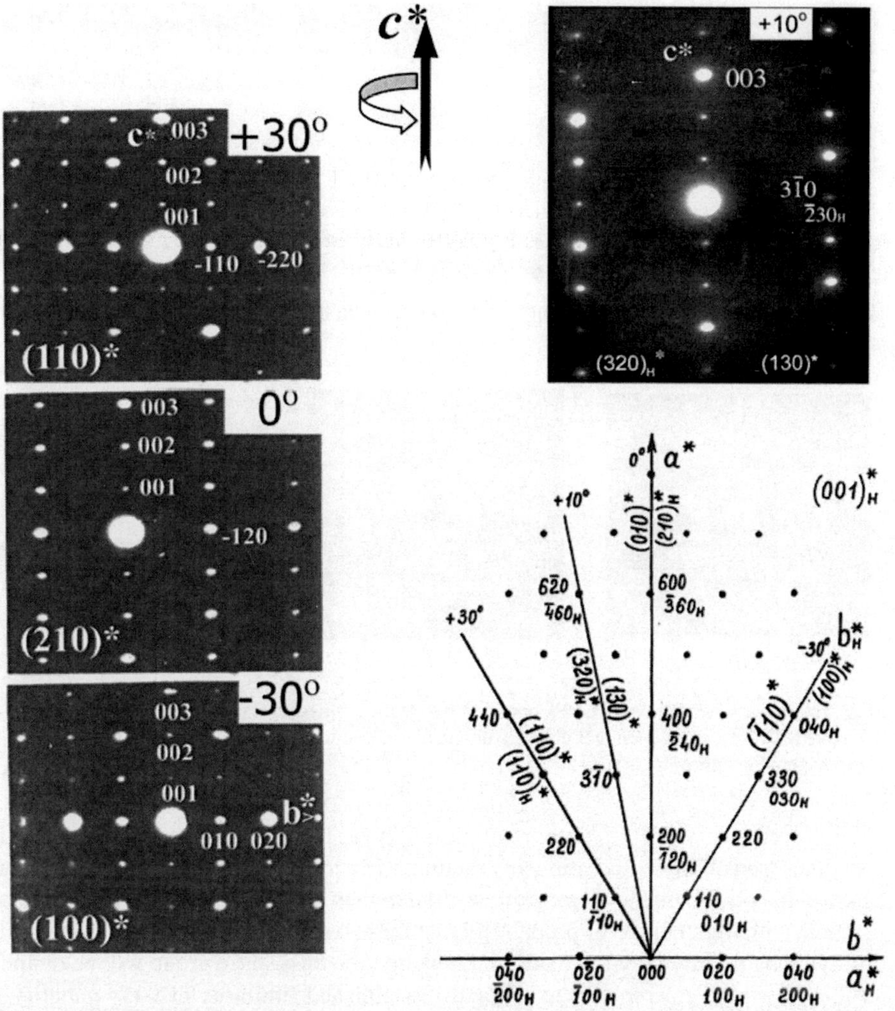

Fig. 1.2 Ningyoite. Series of EDPs obtained when microcrystal is tilted around c* axis, and their decoding scheme: graphical construction of plane, perpendicular to rotation axis. (On right—scheme and MDP—initial indexing carried out in orthorhombic syngony; with index H—indexing in hexagonal one)

reduces number of possible space groups to 4 (from diffraction class D_{6h}): P*622*, P*6mm*, P$\bar{6}$ m2, P$\bar{6}$2 m. Symmetry of intensities arrangement on this SAED-picture is corresponding to (320)* plane, and shows that here symmetry axis of sixth order is inversion one. Consequently, ningyoite structure is characterized by one of two possible symmetry groups' P$\bar{6}$ m2 or P$\bar{6}$2 m.

It should be noted that all reflexes of 000 l type are clearly manifested in all EDPs of ningyoite, however, a sharp increase in intensity *000 l* reflexes with $l = 3n$ is observed, which is characteristic diffraction feature of this phosphate. The intensity

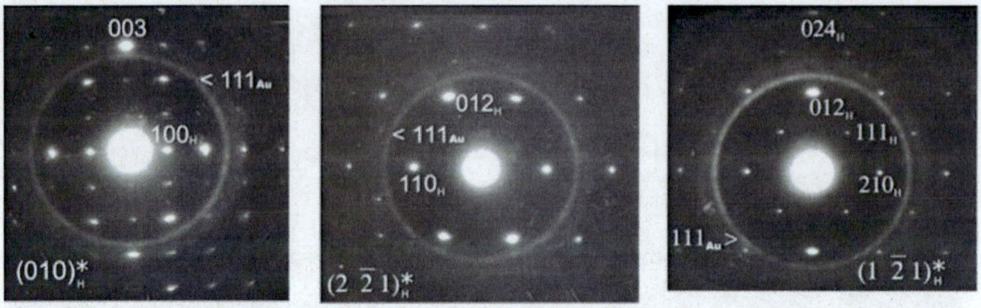

Fig. 1.3 Ningyoite. Point electron diffraction patterns with an "internal" standard (sprayed gold)

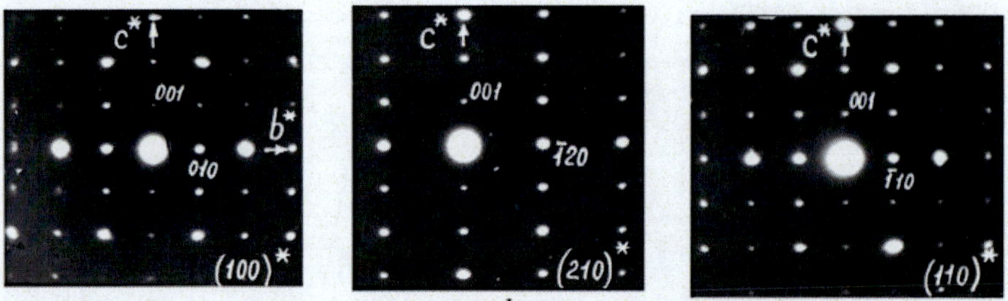

a *b* *c*

Fig. 1.4 Ningyoite. Series of EDPs obtained by rotating microcrystal around elongation axis ***c****: original MDP (**a**); subsequent MDPs were obtained at 30° and 60° (**b**, **c**—respectively). Indexing is in hexagonal syngony

of *000 l* reflexes with $l \neq 3n$ (not multiples of three) is weak. Taking into account that peculiar "sensitivity" (radiation interaction with matter) of electron diffraction is approximately 3 orders of magnitude higher than of X-ray one, it is possible to assume with high degree of probability that these weak reflexes will be absent while studying by traditional x-ray method. Then the mineral's space group will be defined as P6_2*22*, which corresponds to originally established similarity of x-ray patterns of ningyoite and rhabdophane (Muto et al. 1959).

The presence of weak *000 l* reflections can be explained by manifestation of secondary diffraction, which is very likely in case of minerals containing "heavy" elements. In addition, appearance of secondary diffraction is favored by presence of strong reflections on SAED-patterns. Since crystal structure of ningyoite has not yet been determined, conditionally, based on hexagonal symmetry and proximity of parameters with hexagonal rhabdophane ($a = 6.98$, $c = 6.39$ Å), space group of ningyoite is indicated as P6_2*22*, emphasizing belonging of these minerals to one single mineral group (see Table 2.1).

- Decoding *EDPs of minerals with orthorhombic symmetry*.

Point-EDPs of minerals lower-lying syngonies, orthorhombic and monoclinic, have been deciphered by the author usually using reflexes layout schemes.

Geometrical construction (schemes) of theoretical cross sections of reciprocal lattice, for which experimental data were used, has been made.

As an example of determining cell parameters for orthorhombic structures the technique for deciphering SAED-patterns of *vyacheslavite* here is given (Fig. 1.5 and 1.6). Very often, at the normal (perpendicular) orientation of electron beam to specimen, the concrete plane of reciprocal lattice was displayed. This fact allowed us to take it as plane a^*c^*, and further considering plane (010), corresponding it in direct lattice, as basis plane.

The series of electronograms obtained by rotation of individual particles around two mutually orthogonal coordinate axes ***a*** and ***c*** (Fig. 1.5 and 1.6, respectively) in range of angles from −45° to +50° has allowed to establish mutual perpendicularity of all three coordinate axes and to conclude about the orthorhombic syngony of this new mineral. The parameter b was calculated on average value of translation b^* from the distance in the reciprocal space between the basal reflex and spatial reflex corresponding to it at altitude k = 1, 2, 3, … counting multiplicity. The value of parameter b was refined from a number of basal 0 k0 reflections (Fig. 1.5, EDP's tilting angle +8°). The unit cell parameters of this new phosphate, with taking into account X-ray diffractometry data, are (see Chapter 3, sect. 2): $\boldsymbol{a} = 6.96 \pm 0.01$ Å, $\boldsymbol{b} = 9.10 \pm 0.01$ Å, $\boldsymbol{c} = 12.38 \pm 0.01$ Å. Analysis of these EDPs allowed to establishing following extinction laws: for *hkl* reflexes—sum of indices *h* + *k* is even, for *0kl* reflexes—value *k* is even; for *h0l*—values *h* and *k* are even; for *h00*—*h* is even; for *0 k0*—*k* is even; for *00 l*—*l* is even. It follows that spatial group of the mineral is one of: *Cmcm*, $Cmc2_1$ or *C2cm*.

Assuming *orthorhombic symmetry* (and unit cell consequently), author have calculated parameters of uranous phosphate *lermontovite* (possible deviation towards to monoclinic symmetry is permitting). Point EDPs were obtained on different particles, because in normal mode of EM operation particles of this mineral retain ability to diffract no more than 30—40 seconds. The resulting set of SAED-patterns was considered as two series of EDPs that have common reflexes lines: *00 l* and *0 k0*. The characteristic elongated mineral particles shape and placing one of crystallographic axes (***c***) along to elongation of plates (lamellae) makes it possible to estimate mutual arrangement of diffraction patterns obtained. Electron diffraction pattern with orthogonal disposition of reflexes was chosen as initial section corresponding to reciprocal lattice basis plane (100)*. The rest electronograms can have been divided into groups containing common for each group series of reflexes. Thus, electronograms obtained from different particles, but having common reflexes series *00 l* (see Chap. 3, Fig. 3.16–3.18), can be considered as series of different sections obtained by microcrystal rotation around ***c-axis***. Another series of electronograms was considered as result of rotation cross-cutting plane of reciprocal lattice around the ***b-axis*** or close to it. The third elementary-cell parameter, perpendicular to basal plane, was determined by calculation of set electronograms, which have been "tilted" relative to this basal plane. Some asymmetry in intensities arrangement and some deviation from orthogonality in reflexes spacing on electronograms are considered as result of weak deflection of individual mineral lamellar fibers, i.e. not parallel bedding (location) of thinly laminated particles on the

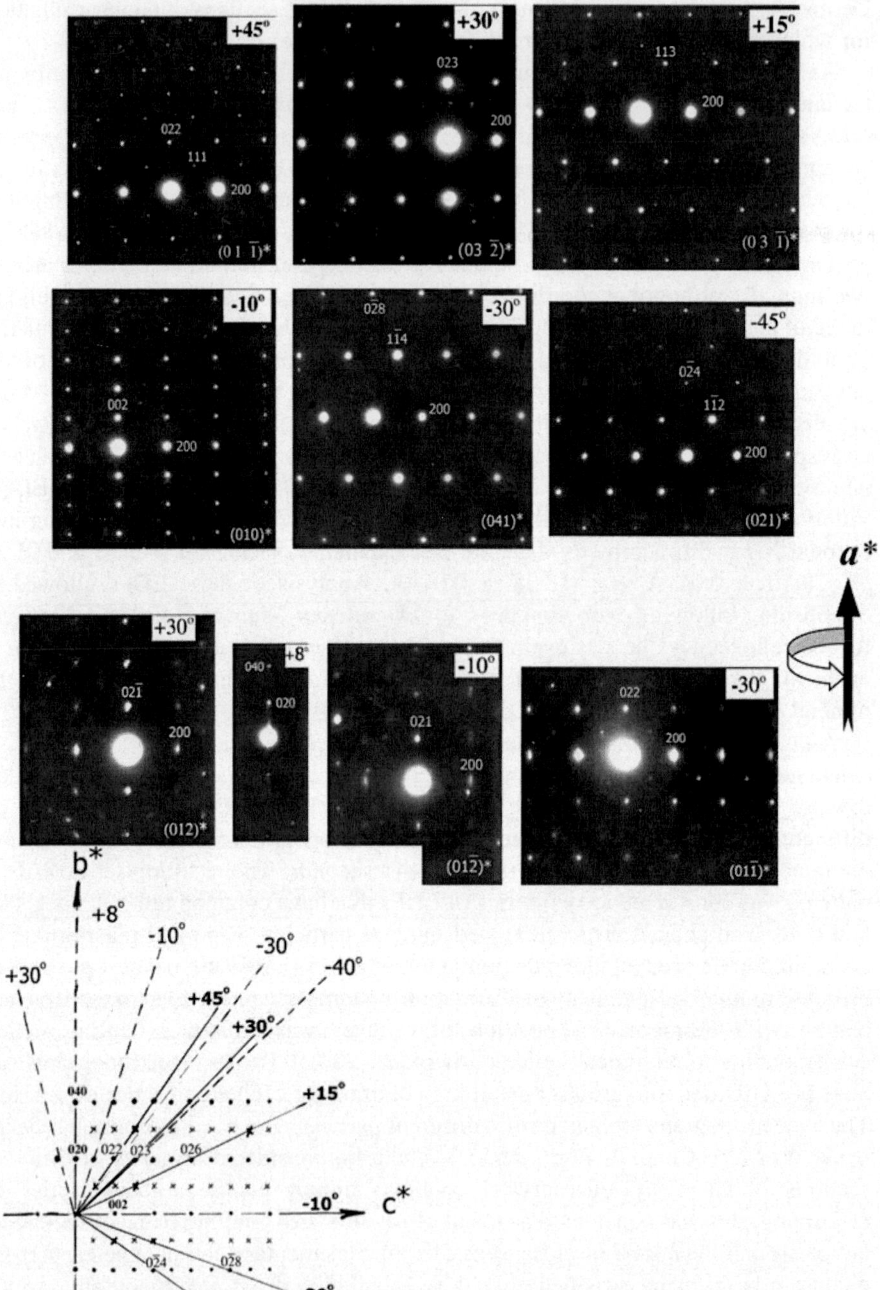

Fig. 1.5 Vyacheslavite. Series of EDPs obtained by tilting microcrystal around ***a**** axis, and scheme for them deciphering. On the scheme, their position in reciprocal space (corresponding to reciprocal lattice cross sections) is marked by solid lines (series top) and dotted lines (series below)

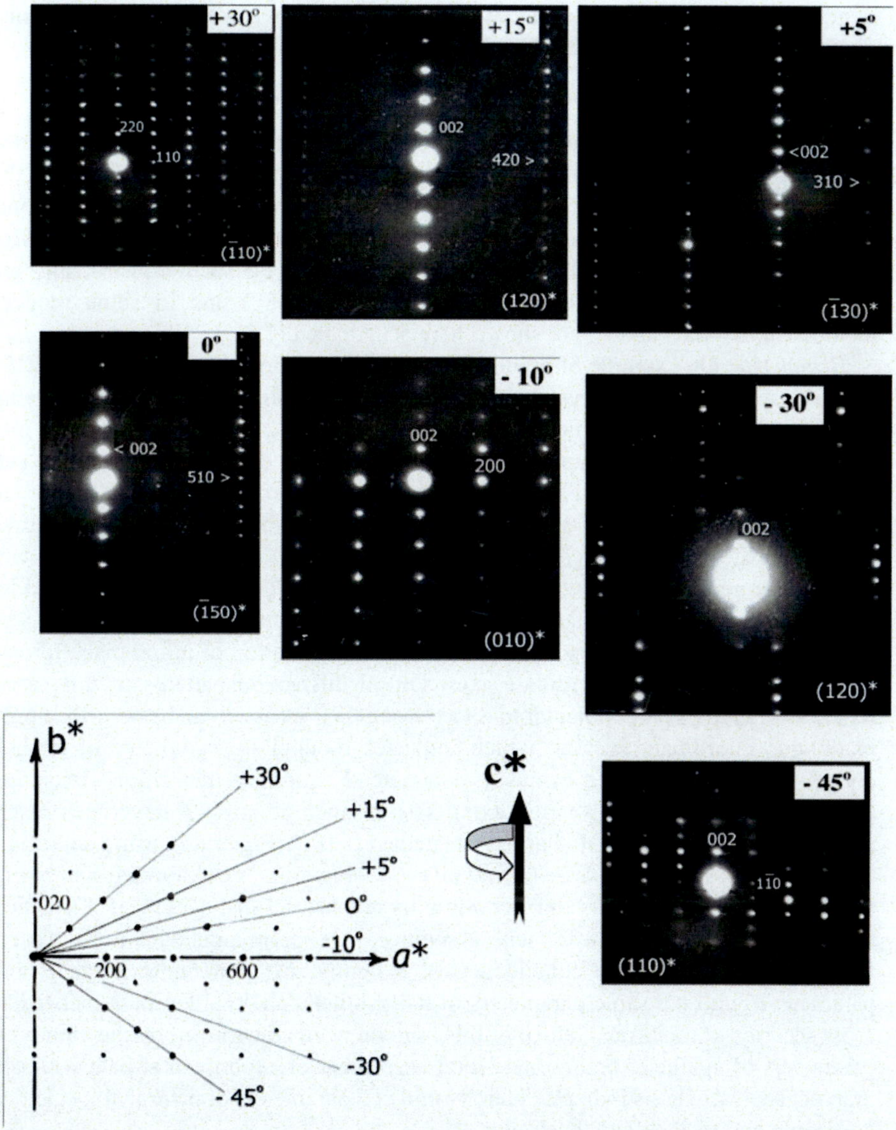

Fig. 1.6 Vyacheslavite. Series of EDPs obtained by tilting microcrystal around ***c**** axis and scheme for them deciphering. The location of these MDPs (i.e. corresponding cross sections in reciprocal space) is shown by solid lines

preparation. Parameters of the mineral have been calculated in assume the elementary cell orthogonality by point SAED-patterns: $a = 8.6 \div 9.8$ Å; $b = 18.6$; $c = 10.1$ Å (Chap. 3).

The author used theoretical constructing of reciprocal lattice plane in study of orthorhombic Y-silicate—kainosite (Nekrasov et al. 1987) and of monoclinic minerals: brannerite (Ivanova et al. 1982), carnotite (Belova et al. 1982), coconinoite (Doynikova and Sidorenko 2006).

- *Decoding the EDPs of minerals with monoclinic symmetry.*

For crystals of more lower-lying symmetry calculation of electronograms becomes much more complicated; and so experimental SAED's data must be supplemented by using additional literary information for theoretical constructions. The author used schematic constructions of assumed coordinate cross sections or that's corresponding to mineral cleavage. Methods of such constructing are described in detail in the author's PhD (Ivanova 1982) and in some modern methodical manuals and so are shown here schematically.

Brannerite. On example of brannerite, methodical scheme for decoding of EDPs for monoclinic symmetry crystal is considering; and obtained data further are being interpret from crystallochemical point of view (Figs. 1.7, 1.8, and 1.9). Taking into account contradictory and limitation of available literary data, definition of unit cell parameters of natural (not calcined) Brannerite was performed by the author on assumption that brannerite structure in its natural (metamict) state is unknown (Ivanova et al. 1982).

Analysis geometry of all obtained SAED-patterns did not allow us to consider that mineral's syngony is higher than monoclinic. Calculation of parameters was carried out from EDP's series, which corresponds to rotation of initial plane (cross-section) around one of coordinate axes. On all diffraction patterns of this series reflexes are in rows perpendicular to tilt axis; and their location can be described only in oblique coordinate system, which indicates monoclinic symmetry of studied crystal. This allowed us to calculate parameter ***b*** value by translation size along the tilt axis (microcrystal rotation axis). The monoclinic angle β have been determined directly from the EDP that reflects plane (1¯10)* (Fig. 1.8d), using additional construction proposed in paper (Zvyagin 1964) (Fig. 1.9a, c). A limited number of electron diffraction patterns did not allow us reconstruct disposition of diffraction nodes in coordinate plane ***a*c**** with assurance, by experimental data alone. Therefore, taking into account coincidence of syngony, of monoclinic angle β and parameter ***b*** with the same parameters of its' synthetic analog, the main theoretical cross sections of reciprocal lattice of this mineral were constructed by the author on parameters of synthetic brannerite. Good agreement of experimental data with our graphical constructions indicates identity unit cell parameters of natural non calcined brannerite and its synthetic analogue.

Thus, by using electron diffraction, structural characteristics of metamict brunnerite in its natural state were obtained (usually for such studies in mineralogical practice preliminary calcination of sample is used). Parameters of nature brunnerite defined by SAED method: $\boldsymbol{a} = 9.8$, $\boldsymbol{b} = 3.8$, $\boldsymbol{c} = 6.7$ Å; $\beta \approx 119°$—correspond to monoclinic unit cell in synthetic brannerite and cell in thorium titanate isostructural to it. Evidence of structural similarity between brannerite and synthetic $ThTi_2O_6$ is based on coincidence as of syngony, monoclinic angle, parameter ***b***, so on intensities of basal *00 l* diffraction reflexes (visual comparison) (Ivanova et al. 1982).

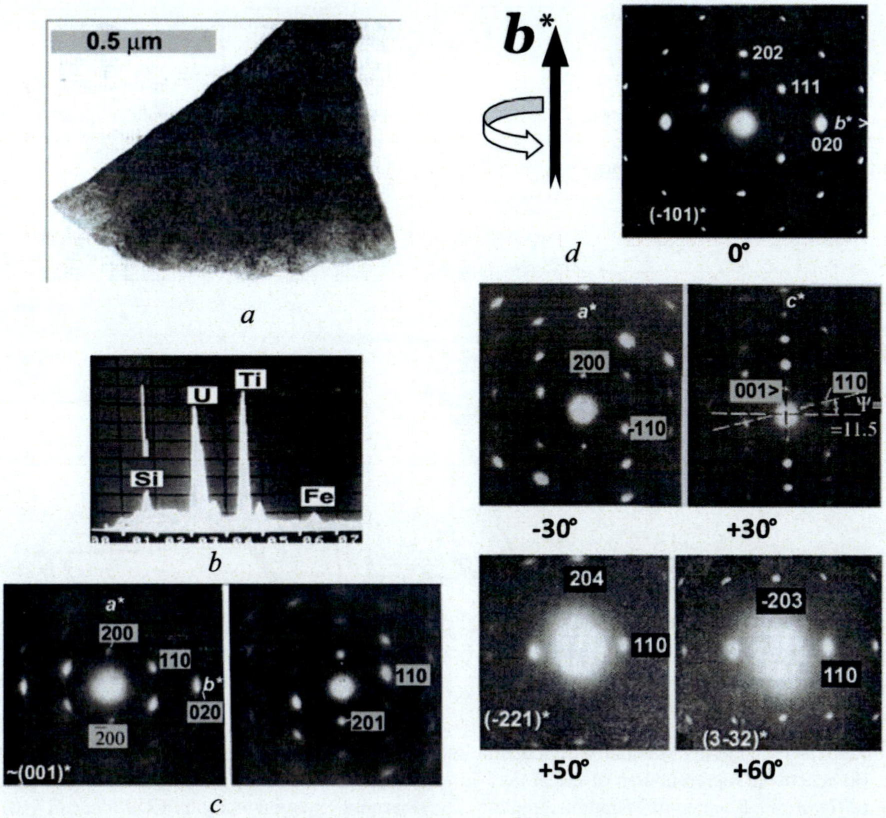

Fig. 1.7 Brannerite. (**a**) mineral particle TEM image; (**b**) its composition ED-spectrum; (**c**) EDPs reflecting brannerite reciprocal lattice cross sections close to (001) * and $(11\bar{2})$ *, respectively; **d**—set of EDPs obtained at different angles of the particle tilt around ***b**** axis, that corresponding to nodal planes, respectively: 0° – close to $(10\bar{1})^*$; −30° – $(001)^*$; +30° – $(1\bar{1}0)^*$; +50° – $(2\bar{2}1)^*$; +60° – $(3\bar{3}2)^*$

The similarity of structural characteristics and crystal-chemical properties of U^{+4} and Th^{+4} allows us to speak of the isostructural nature of synthetic UTi_2O_6 and $ThTi_2O_6$, and analyze brunnerite structure using known data on its synthetic analogs.

The structure of synthetic $ThTi_2O_6$ (Fig. 1.9) is composed of stepped layers of identical TiO_6 octahedrons (Ruh and Wadsley 1966). In each step of a layer of octahedrons, which connected in pairs with sharing an edge, these pairs are connected by vertices along the b-axis.

In each stair-step of a layer, the octahedrons are connected in pairs by common edge, and these pairs are connected by vertices along ***b*** axis. The adjacent steps of a layer are displaced relative to each other along ***b*** axis by half the cell (½ of octahedron diagonal). Steps are interconnected in such a way that each octahedron of one stage is connected by edges with two octahedrons of other step. This Ti-O layer can also be represented as an association of the Brookite type chains (thick line

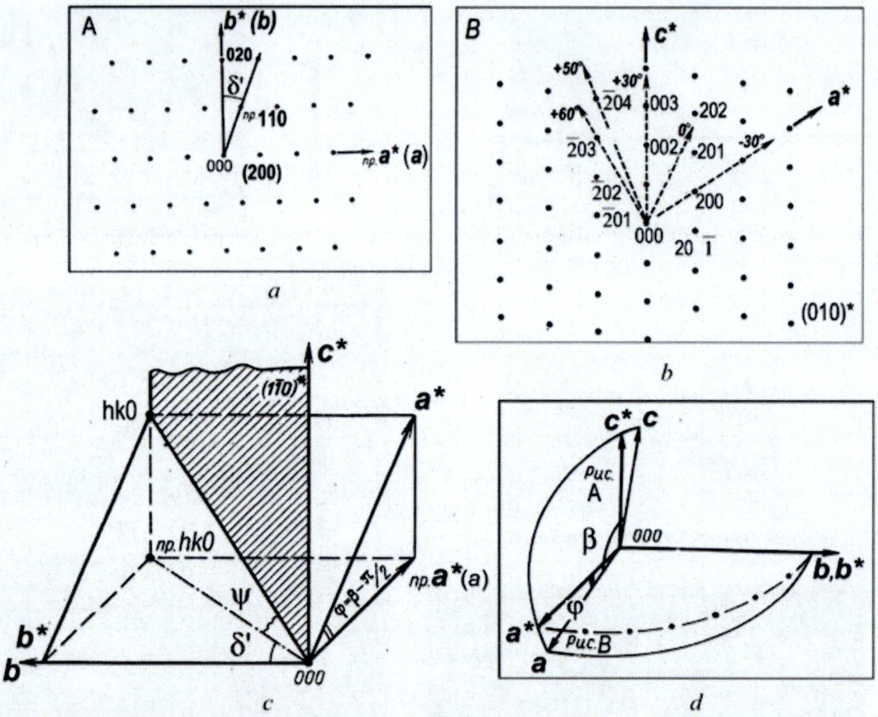

Fig. 1.8 Brannerite. Methodical scheme for interpretation electron patterns of monoclinic mineral: (**a**) schematic representation of the projection of (001) * plane of reciprocal lattice onto the plane (001) of direct lattice of brannerite, angle δ` are measured experimentally on EDP plane (1¯10)* (see Fig. 1.8d, +30°); (**b**) schematic image of nodal plane (010)* of brannerite reciprocal lattice (according to the literature), dotted lines show position of electron diffraction patterns crossing it at right angles (set of rotation around ***b**** axis); (**c**) scheme for determining the monoclinicity angle from section (1¯10)* according to (Zvyagin 1964); (**d**) relative position of coordinate axes of direct and reciprocal lattice in monoclinic syngony, showing relationship of schematic constructions ***a*** and **b**

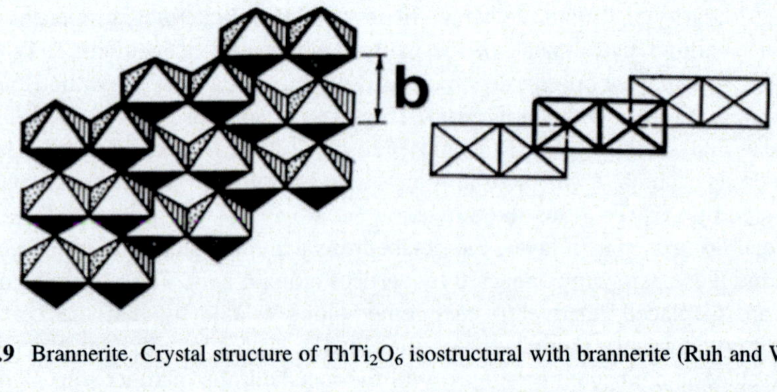

Fig. 1.9 Brannerite. Crystal structure of $ThTi_2O_6$ isostructural with brannerite (Ruh and Wadsley 1966)

in structure diagram, Fig. 1.9). In the projection to (010) plane, the layer looks like a step chain of duplicate (twin) octahedrons (Fig. 1.9). The layers are connected through Th atoms located in the interlayer octahedral positions.

For designate of brannerite structural motif Yu.A. Pyatenko et al. (1976) has introduced the principle of O-tops coordination by Ti atoms; that is resulted to structural symbol 1(1) – 2(2) – 3(3). Such record indicates that from 6 vertices of TiO_6 octahedron one vertex (O_1) is not generalized, two O-vertices (O_2) belong simultaneously to two octahedral, and three remaining (O_3) belong to three octahedral at the same time. Therefore, for each Ti atom in the structure there are 3 O atoms, and layered Ti-O-motif of brunnerite structure is characterized by radical $[TiO_3]^{-2}$. Thus, it was shown that name "titanate" of uranium, corresponding to nomenclature and more accurately reflects crystallochemical nature this mineral than notion "complex oxide" of uranium and titanium, which is widely used in literature in describing brannerite.

Carnotite. The construction of theoretical cross section (010)* of reciprocal lattice was used by the author in deciphering electron diffraction patterns of monocline carnotite (Fig. 1.10). Object of study was cryptocrystalline black-colored formations unusual for natural carnotite (Belova et al. 1982). Since light optics did not allow us carrying out diagnostic, the mineral was studied as a new one, i.e. unknown before. The plate microcrystals of this mineral are characterized by a point EDPs that corresponds to the cleavage planes (001). The calculation was carried out by a series of point SAED-patterns obtained at tilt around axis *b* (*b**) of monoclinic microcrystal of carnotite. Because of absence of cross-sections, from which monocline angle could be determined reliably, the author used theoretical construction scheme of $(010)^*$ plane perpendicular to tilt axis (Fig. 1.10). Taking into account the dehydration of minerals under vacuum conditions of an electron microscope, the theoretical constructing have been based on known parameters of anhydrous synthetic analogue of carnotite. Then, decoding procedure was aimed at elucidating the degree of correspondence of experimental data to model structure of anhydrous synthetic of carnotite analogue. The agreement between the calculated and experimentally obtained interplanar distances d_{hkl}, as well as the agreement of experimental and theoretical angles between successive (upon tilting) cross sections of reciprocal lattice, was established. This made it possible to reliably diagnose carnotite. The values determined by SAED-patterns: $d_{010} = 8.4$, $d_{100} = 10.5$, $d_{001} = 6.4/l$ Å, $\beta \approx 114°$—served as basis for determining mineral in its cryptocrystalline modification.

Methodical difficulty in graphical method for determining parameters is volume extention of knots reciprocal lattice. The plane corresponding to electron diffraction pattern can cross diffraction knot in some angular range; consequently, the calculated value of parameter will have some spread of values. For greater certainty, it is desirable to determine parameter value from reflexes lying on corresponding coordinate axis (so-called basal reflexes). At present, computer-based construction of geometric schemes for location of electronic diffraction reflections is widely used. Now it is also possible to simulate electron diffraction patterns, i.e. given planes of

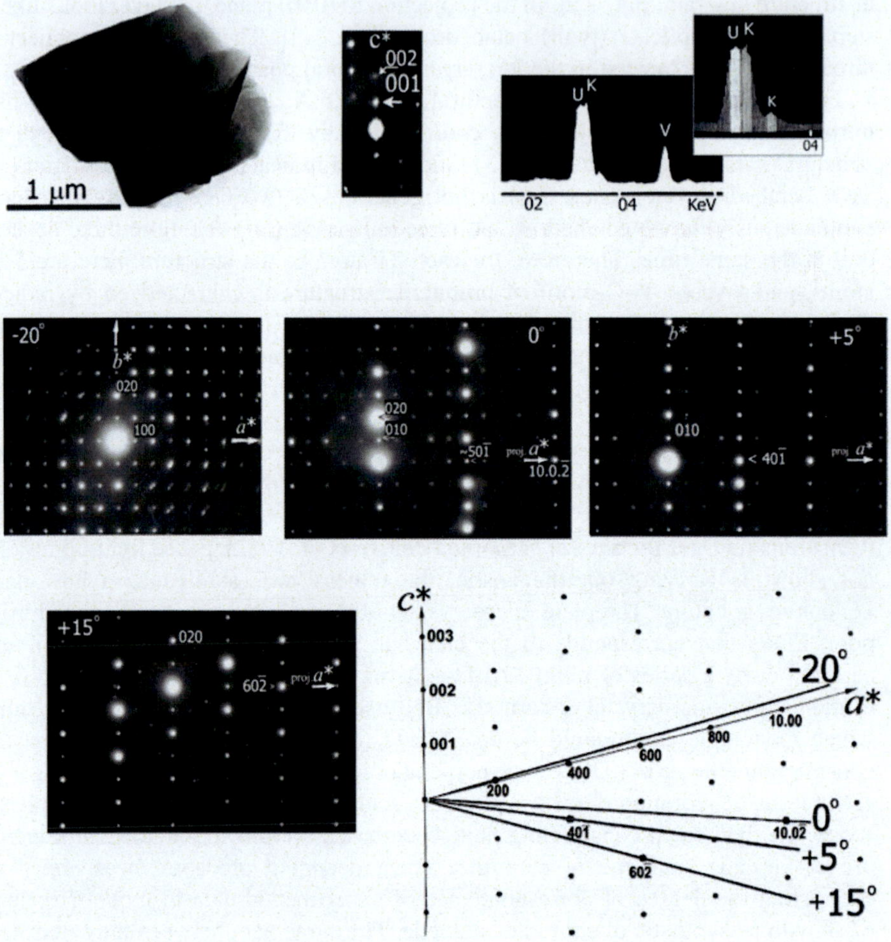

Fig. 1.10 Carnotite. Methodical scheme of monoclinic mineral ATEM-research (Ivanova et al. 1982): (**a**) plate particle TEM image; (**b**) basal reflexes from its flexure; (**c**) ED-spectrum and its fragment; (**d**) set of electron diffraction patterns from the particle, obtained by its tilting on +5°, +15°, −20 °, that corresponding to rotation crosscutting section (in reciprocal space) around axis *b** (*b*); d—graphical construction of (010)* plane for interpretation this set of electron diffraction patterns

the reciprocal lattice of crystal structures, taking into account the reflection intensities (weight), for the assumed structure.

- *Decoding point EDPs of minerals with large parameters.*

In electron diffraction study minerals with "large" (>10 Å) parameters of unit cell and without pronounced cleavage, it is very difficult to obtain EDPs that strictly correspond to definite planes of reciprocal lattice. For such minerals, even with axes orthogonality, symmetry of point-like EDPs is usually skew-angular, as a consequence of random orientation. A procedure which in this case facilitates diagnostics of unknown phase was used by the author in studying mineral of orthorhombic syngony. The mineral was studied, for which at first only basic elements of composition (Si, Y) were known. From a large series of EDPs obtained at different tilts of the microcrystal, only few MDCs could be attributed to orthogonal symmetry. Therefore, at initial stage of decoding, qualitative assessment of all SAED-patterns was carried out. For further processing, EDPs with symmetrical geometry of reflexes net and with symmetric intensity distribution were selected. Interplanar distances *d* of reflections, close to center, and visual rating of intensities were calculated from these electronograms. This set of d_{hkl} was used for preliminary qualitative diagnosis of the mineral by X-ray tables (ASTM etc). On this basis among a number of compounds, that are similar in composition, presumably have been selected orthorhombic Y-silicate kainosite. At second stage, in view of X-ray data, proximate decryption has been performed: indexing, calculation of angles between reflexes, comparison of calculated and experimental angles between ED-patterns. One-to-one correspondence of each obtained diffraction pattern to kainosite crystal lattice was established. The unit cell parameters were determined: $\boldsymbol{a} = 13.13 \pm 0.02$, $\boldsymbol{b} = 14.25 \pm 0.04$, $\boldsymbol{c} = 6.71 \pm 0.03$ Å. The values given in X-ray tables (ASTM) were refined directly on series of *h00* reflections; it is ***a*** parameter and indices of reflexes kainosite. In this case, only TEM + SAED allowed to diagnoses this Y-mineral, first found in the Russia (Nekrasov et al. 1987).

Some methodological difficulties of uranium minerals study

1. Diagnostics of uranium minerals by electron diffraction methods is complicated by phenomena of *secondary diffraction*, which is caused by incoherent scattering of rays in crystal. This is expressed in appearance on diffraction patterns the reflexes forbidden by extinction law. The main reason for secondary diffraction manifestation is existence heavy element—uranium—into mineral's structure. This is in much greater extent than for other minerals (at similar particle thicknesses) enhances effect of dynamic scattering. It must be noted that the newest method PED (precession electron diffraction) act so that such dynamical diffraction effects, that forming secondary diffraction, may be effectively suppressed if the crystallite is sufficiently thin, especially at large precession angles (Moeck et al. 2011).
2. The *phase transformation* a number of mineral phases (usually layered water-containing ones) *under electron beam* also complicates diagnostics of uranium minerals. In article (Ivanova 1982), phenomenon phase transformation uranyl phosphate (uranium mica) into uranium oxide under vacuum of electron microscope was studied. Earlier, all cases of solid-phase transformations under electron microscope conditions described in literature was related only for minerals of groups oxides and hydroxides Fe, Al, Ca, Si, etc. At various stages of gradual transformation of meta-autunite into uraninite elemental composition of the particle irradiated by electron beam was monitored. It has been established that with complete disappearance of dot reflexes of U-mica, only annular diffraction pattern of uraninite gradually appears and remains. The irradiated particles retain

contours and element composition of original U-mica, although they are essentially aggregate of fine uraninite crystallites. Elements remaining in such "uraninite" particles from initial composition of U-mica (P, As, Ca, etc.) can be removed by subsequent stronger irradiation (which leads to melting substance within particle boundaries) (Ivanova 1982; Belova et al. 1981).

3. Features of ED-patterns of layered minerals that are characteristic of *minerals with perfect cleavage*, as uranium mica and layered silicates can also be attributed to methodological difficulties. As a result of suspension preparation, textured specimen with predominant orientation of micro-plates (on substrate) is obtaining. In this case, set of annular diffraction reflexes, which is due to azimuthally turning of micro particles, will not be complete. Some reflexes, permitted by extinction law will be absent. It is ones that characterize structure periodicity in direction of falling rays, i.e. along normal to stratification (with values $l \neq 0$).

In such cases, technique procedure of ED-patterns shooting from bends along micro-plates edges is irreplaceable (Gorshkov 1970). It is used when almost all particles in preparation are perpendicular to the electron beam, and it is impossible to obtain diffraction information on structure layer periodicity in usual way. This method, obtaining *basal reflexes from curved edge*, is still widely in demand for study poorly crystallized minerals with layered structure. If interlayer water is present in mineral, combination of electron and X-ray diffractometry is used. Such examples of using analysis of basal reflexes intensities for studying structures weakly crystallized minerals are numerous works of Russian scientists V.A. Drits, A.I. Gorshkov et al., which devoted to study Fe-Mn-minerals. This technique was used by the author in studying black carnotite (vanadium U-mica).

1.1.2 Composition Analysis of Dispersed Minerals in EM

Current level research involves equipping both TEM and SEM by spectrometers for chemical composition analysis. In EM mainly energy-dispersive spectrometers (EDSs) with dispersion of registered radiation by energy (Kevex, Link, etc.) are used. EDS's advantages are possibility observation on screen entire elements spectrum that make up mineral composition and analysis rapidity. Time quantitative analysis is $50 \div 100$ seconds, and qualitative one—several seconds. It is possible to register whole elements range from Be (for modern detectors) to U. Detection limit is 0.1% wt for heavy, and 0.01% for light elements. Image resolution in BSE mode (back-scattering electrons) is 0.1 Z; here contrast act as a function of the average atomic number Z.

In TEM, semi-quantitative analysis particles with the size of about 1 μm (at best, micron ones) can be carried out. Here, for quantitative analysis, the lower bound of analyzed particles size is determined by sensitivity and by probe beam diameter.

When massive samples composition studying by EDS, analysis localness (1÷7 μm^3) depends on the average atomic number (Z) of object/sample.

All techniques of quantitative microanalysis are based on the possibility of using standards. There are possibility of qualitative / quantitative analysis at the point and by area is possible, as well as analysis of distribution of elements along the profile or by area. The analysis "at a point" assumes analysis of certain volume, which occupies excitation region from falling electron beam. Depth of area excitation varies within range of n·μm^3, depending on mineral elements composition. In the case of fine mineral particles (analysis in ATEM), it is obvious that the excited volume is less than in the bulk sample (analysis in ASEM). Calculation of analysis is based on determining the ratio between X-ray radiation intensities of element in the sample and in standard.

The image in X-ray characteristic radiation of an element gives qualitative information about relative distribution of this element within scanning area. In this case, the difference in the content of an element for separate segments is recorded as difference in the average number of points per unit area. Visually we can distinguish only significant changes in the content of element on this image. Principles of electron-probe analysis, the technique of X-ray spectral measurements and their interpretation are considered in detail in a number of methodic works (Gouldstein and Yakovits 1978; Reed 1979, etc.).

For mineralogical studies analysis of composition in ATEM is the most informative in the case of smallest particles of matter, since it complements the diffraction characteristics. The EDS analysis in combination with electron diffraction (SAED) allows obtaining a set of crystallochemical data for the same micron particle that necessary for reliable diagnostics.

Elemental composition analysis in ATEM is carried out on the thin homogeneous mineral particles (up to 0.1 μm), which lie separately, isolated from each other, on the suspension substrate. The suspension preparat should be sufficiently sparse to exclude the influence of neighboring particles onto spectral characteristics of composition. The analysis area is determined either by beam diameter (from ~1 μm to n Å in modern models) or by the particle size (if it is smaller than probe diameter). The analysis area coincides with SAED-research zone (Mokhov 1987) and can reach n.10 Å. Quantitative analysis realization for thin objects (at transmission) methodically is very different from analysis of massive/bulk samples, and therefore the method itself was developed much later (Cliff and Lorimer 1975; Mokhov and Tsepin 1979, 1980). When minerals study, homogeneous etalons are usually chosen, which average atomic number (by elements entering into them) is closest to that for investigated objects. To perform quantitative analysis of thin mineral particles, as standard was used a mineral with the same elements set and similar in content as investigated object (Mokhov 1986). This allowed avoiding the calculation of various corrections and, through some transition coefficients, to determine relative mass content of elements in object under study.

When analyzing thin particles by transmission, quantitative calculation is possible only under conditions normalization of characteristic spectrum intensities; i.e. resulting elemental composition is always normalized to 100%. Quantitative analysis in ATEM using EDS is the most local of all currently existing; however, it requires special experimental research and significant time-consuming costs (Mokhov and Tsepin 1980).

Although the sensitivity ED spectrometers is not high enough, they are indispensable in the study of dispersed mineral substance formations. To study loose uranium ores, the efficiency of the ATEM SAED complex is very high. Unfortunately such equipment has not yet been widely used in mineralogy. The reason is, first of all, complexity of experiment. Each of methods (SAED, high resolution, phase contrast, vacuum decoration and quantitative composition analysis) requires individual approach, both to conducting experiment, and to processing results.

Methodical notes.

Long-term practices of working with analytical transmission electron microscope allowed to generalize some methodical features of disperse uranium minerals investigation.

1. The manifestation effect of secondary diffraction, in most cases, is due to appearance reflexes, forbidden by extinction laws. So, for example, in the case of coffinite (sp. group $I4_1/amd$); forbidden reflexes *002* may appear, and in the case of ningyoite (sp. group $P6_222$)—reflexes *000 l* with $l \neq 3\mathbf{n}$. Weak of these reflexes intensity in comparison with all other ED-reflexes, as well as the presence of heavy element (uranium) in mineral structure, allows to linking their manifestation with the effect of secondary diffraction. More rarely, secondary diffraction is manifested in the "leveling" (smoothing) of intensities on diffraction pattern: weakening of strong and strengthening of weak reflexes. Effects of secondary diffraction disappear at a small tilt (n°) of microcrystal.
2. During EDS analysis, a particle under study undergoes a long (up to 3 min) beam thermal impact. If the mineral is structurally unstable (its structure is weakened by processes of dissolution, substitution), diffraction pattern may degradates during operating, and through to it's disappear completely. Therefore in ATEM, it is advisable to carry out composition analysis after getting SAED patterns on this mineral particle under study.
3. The poorly crystallized X-ray-amorphous uraninite is often present in composition of exogenous uranium ores. This ore phase will be "lost" when determining mineral composition with the help of X-ray alone. Such mistake was made earlier when ore composition of Chu-Sarysuy ore province have been diagnosed by X-ray as mono-coffinite one. For a complete diagnosis mineral composition of finely dispersed exogenous ores the complex of analytical methods ATEM +EDS + SAED is indispensable.
4. Experience of uranium minerals studying shows that during preparation by suspension, well-crystallized varieties of uranium oxides significantly much faster drop to glass-tube bottom than less crystallized or other uranium minerals. In order to exclude diagnostic errors of "underestimation" (not to "miss" one of the phases), it is necessary to investigate both sediment and suspended material of the suspension.

1.2 Scanning Electron Microscopy Possibilities in Minerals Study

Current level of geological and mineralogical research requires using of scanning electron microscopy (SEM) and its analytical capabilities. This method is used today practically in all fields of science and industry, in biology, biomedicine, paleontology, physics, materials science, etc. The method is designed to study massive samples.

To obtain an image of the object surface with high spatial resolution, SEM uses the reflected and scattered rays produced by interaction of electron beam with object. The method is based on television principle of surface scanning by electron beam. The interaction of electron beam with solid sample causes different types of radiation and, accordingly, there are various methods for investigating by scanning EM (composition analysis, phase and magnetic contrast, cathodoluminescence, etc.).

In mineralogy ASEM is most often used for volume, not polished objects, when sample integrity is not disrupted by preparation. Minerals are observed in their natural relationship, with increasing of 20–150 thousand and resolution of 100–150 Å (in modern models—10 Å). Morphological features, details of micro relief, interrelations of mineral phases are under studied. In the case of relief samples qualitative composition analysis for phase diagnostics is used.

Quantitative analysis of composition of a mineral substance is possible only in the case of polished sample surface (polished section). Mineral samples are often poorly conductive. Therefore, preliminary deposition of thin conductive material film onto sample is necessary to provide static charge running-off, which caused by electron flow. Usually graphite, aluminum or gold is sprayed onto surface.

Diagnosis of mineral phases can be ambiguous when using only quantitative analysis, for example, in the case of polymorphic modifications or close in composition minerals. When disperse (micron-sized) mineral matter is investigated by ASEM methods, the question of result synonymity always remains open (even on polished samples). Namely, to what extent does get composition correspond to phase under study, since the surrounding matrix contribution cannot be completely excluded.

Recently, in the Scanning EM to diagnose crystal structures method electron backscatter diffraction (EBSD) focused mainly on materials sciences is being used. This method (also known as Kikuchi diffraction) was developed mainly for metals and alloys, and so minerals; is used in materials engineering to solve applied problems. EBSD patterns can be used for indexing, determining syngony and crystal orientations, studying defects and intergranular boundaries, etc. This method is of significant interest for the characterization of nanocrystalline materials and nanostructures. There is equipment for SEM to getting kikuchi-diffraction patterns on thin samples (so called transmission t-EBSD), but such thinned sample preparation processes on minerals are very demanding and time-consuming. Diagnostics of an object (mineral) is performed indirectly by selecting the most nearest phases, similar by interplane distances and its' intensities, from a wide list of the database.

Specialists suggest are considering t-EBSD as a complementary technique to both EBSD and TEM. In case of micron/submicron mineral objects TEM's diffraction methods again come to forefront in diagnostic procedure.

The comparing analytical capabilities of analysis methods (different analyzers) are interesting and important for mineralogical studies. The principle of implementation of elemental analysis is the same: intensities of X-ray characteristic radiation that individual for each element are measured in this individual range of wavelengths or energies. Methodical aspects of qualitative and quantitative analysis are described in detail in the literature (Reed 1979, etc.) and in a lot of methodical manuals.

The possibility of composition analysis of minerals in TEM and micro probe is caused to the different types of spectrometers. The operation principles of two types of spectrometers and their comparison are discussed in detail in many methodical publications (briefly—in a number of advertising brochures of analytical equipment). The following brief discussion refers to the study of massive polished mineral samples. The limits of the RSMA and ASEM methods are erased when modern micro probe models are equipped with both types of spectrometers.

According to the manner of radiation recording, the spectrometers are divided into: *wave-dispersion* (WDS) and *energy-dispersion* spectrometers (EDS). In mineralogical EM—investigations are used WDS (micro probe) and EDS/the analytic possibilities of SEM method to the solution of mineralogy problems are briefly described, for example, on the site of IGEM RAS (www.igem.ru).

Both types of spectrometers are able to register all elements in the range from Be to U. In the case of EDS there is possibility of observing simultaneously the full set (spectrum) of elements entering the mineral composition, this procedure is impossible for the WDS. EDS favorably differs by the express quantitative analysis (50–100 seconds), but for this analysis the requirements to the quality of polishing are higher. EDS is significantly inferior to wave spectrometer in sensitivity due to a lower peak-to-background ratio. The energy resolution of EDS is significantly worse than that of VDS, which complicates analysis at low element concentrations and when overlapping peaks. For WDS, a full spectrum analysis usually takes 5 $\div$ 20 minutes; a high spectral dispersion of wave spectrometers minimizes the overlap of the spectral lines to each other.

These two types of spectrometers have a close sensitivity only in those regions of the spectrum where the peaks of the elements (spectral lines) are easily separated. Only in these spectrum regions the comparison of the detection limits in quantitative analysis of the elements is correct: for EDS up to n $\div$ 0.n wt. %, for WDS usually 0.n $\div$ 0.0n wt. %. Lower currents in the EM (by one order compared to the micro probe) less heat the sample and, therefore, allow studying a wider range of mineralogical objects (that "burn" under a beam, water-containing, etc.).

1.3 Objects of Research, Equipment

The objects of this study were mainly ores of uranium exogenous deposits in a loose sedimentary cover: samples from secondary enrichment zones (redox mineral zonality), from weathering crusts of endogenous uranium deposits (L.N. Belova's samples), and the ningyoite ores of paleo-valley-type deposits of the Khiagdinsky Ore Field (samples of N.N. Tarasov). The uranium-ore samples were usually poly-component powdered mixture. Mineral composition of uranium ore substance was studied by TEM and AEM; when the last—radiography has been used as preview method. Diagnostics of ore minerals was carried out by complex methods of TEM and SEM (SAED, EDS, BSE-imaging). Samples were studied in different EM preparations: handpicked individual grains, transparent polished sections, and polished sections.

The author carried out a crystallochemical study of uranium minerals in loose fine-dispersed mineral mixtures, based on the SAED method, and phase diagnostics of ore matter in the samples. The work was carried out mainly on ATEM JEM-100C, Jeol, Japan (1969). The accelerating voltage is 100 kV. The image resolution is ~4.5 Å. Equipment with a goniometer provides azimuthally rotation (in the plane of the drug) by $\pm 180°$ and a tilt of $\pm 60°$. Selector diaphragm with a diameter of ≤ 1 μm (to isolate the studied region) provides the local diffraction studies. EDS—Kevex-5100 (USA) and raster equipment ASID-4 (for samples $\approx 4 \times 6 \times 8$ mm) with resolution up to 15 Å in transmission mode, in secondary electrons ~30 Å.

More recent studies were conducted on the ASEM JSM-5300, Jeol, Japan, (1992); resolution of the 3D image in SE mode ~ 40 Å. EDS—Link ISIS, Oxford, England, (1993). The analytical studies of bulk samples have been performed as by qualitatively (when it relief), and quantitatively (when it polishing). Uranium ores of paleo-valley deposits were studied at JSM-5610LV, Japan. Accelerating voltage 25 kV; 100 kV; BSE images resolution ~400 Å (reflected electrons mode). ED spectrometers—INCA-Energy 450 and Aztec, Oxford, UK. Quantitative analysis elements range is from Na to U. The selections individual grains of sandstone and the thin polished samples, that was prepared from loose material by epoxy glue cementing, have been studied.

1.4 Electron Microscopy Contribution to Uranium Mineralogy (some History)

Significant progress in uranium mineralogy study since the 70s is largely due to the use of EM in the daily practice of geological researches. Successes in the study of complex uranium mineralization, achieved in EM-laboratories of scientific institutes Moscow (IGEM RAS, VIMS), stimulated the creation of electron microscopic divisions in industrial organizations in the late 1970s. Uranium ores are often difficult objects for traditional mineralogical methods of diagnostics (optical,

X-ray, microprobe analysis, etc.) due to fine dispersity or metamict state of the substance. The study of fine-dispersed uranous mineralization is often hampered by the extremely small amount of ore phase. Significant contribution of electron microscopy to the uranium mineralogy must be underline here.

The first experiments of EM study of uranium minerals were related to the study of particles morphology (Gritsaenko et al. 1961; Belova et al. 1976). Since the 1970s, the contrast radiography technique EMAR (electron microscopic auto-radiography) has served to visualize the distribution of radioactive elements in a sample. Thus, it was fixed that the major part of RAE in hyalite (variety of opal) is distributed in the mineral-owner unevenly, and the element's own mineral form is represented by oxides of U^{4+} and Th, by U^{6+} phosphates or amorphous formations (Sidorenko et al. 1986). Here, the methods of preparing weakly cemented uranium ores for fixing a loose matrix and porous samples have been described in detail. Using the SAED method, the structure of tiny crystallites of coffinite (R factor = 0.04) was determined. Using this method, the authors convincingly, as it seems, have put an end to the long-standing scientific debate about the primitivization of uranium oxide face-centered cubic (FCC) lattice due to oxidation. It has been proved that UO_2 destruction proceeds without the primitive intermediate phase formation, and the reason for conclusion about decrease of the symmetry of the unit cell is the erroneous indexation of ring SAED-pictures.

The study of brannerite (uranium titanate) with the help of replicas revealed areas with clearly expressed crystallites (up to 2–4 μm) onto the cleavage surface of its metamict crystals (Dubinchuk and Belyaevskaya 1975). In SAED investigations of brannerite by the author (Ivanova et al. 1982) was confirmed the conclusion about the possibility to find particles brannerite that demonstrate the monocrystal microdiffraction pictures. Replicas method is widely used up to 1990s for explore the micro morphology of uranium—and uranium-bearing minerals. It allows you to receive inaccessible to other methods information about the uranium mineral formation. For example, TEM study of ordinary ores of infiltration deposit not only revealed uranium mineralization form, but also allowed to obtain genetic information about the nature of uranium accumulation in such ore deposits (Dubinchuk et al. 1990). The signs of ore minerals dissolution in ores of the stratum-infiltration deposits, revealed with the help of replicas, convincingly testify to the processes of redistribution of uranium mineralization, which fully corresponds to the theory of infiltration ore formation and data of other methods (radiochemical, isotope).

A vivid example of the effectiveness of the use of complex of EM methods (SAED + EDS) in combination with other structural methods (electron diffraction, X-ray, thermal analysis, IR spectroscopy) is a comparative study of urgites and hydronasturanes (Dubinchuk et al. 1981). It has been established that hydronasturanes are an intermediate form of the change (oxidation) of uraninite up to its complete amorphization, resulting in the formation of urgite. The authors made a conclusion important for understanding the hypergene mineralogy of uranium: the formation of crystalline uranyl-hydroxides is preceded by the amorphization of the primary uranium oxides.

In the article (Dubinchuk et al. 1990) the long-term experience of EM studying dispersed exogenous uranium ores from various deposits (uranium-coal, “sandstone type” and paleo-riverbed), which are united by the common nature of ore-forming infiltration processes, was summarized. The similarity of the forms of uranium ore minerals separation and the similarity of their composition are revealed, that shows the fundamental similarity and commonality ore mineral formation—genetic uniformity. Non-mineral (sorbed) form of uranium concentration, bound to the surface of clay minerals and feldspar, was clearly fixed.

In recent times the interest in uranium mineralogy has revived from a new point of view. After extensive research in the 1940s and 1950s, interest in uraninite again increased as for it is an analogue of UO_2 in spent nuclear fuel. So the study of the early stages of uraninite oxidation is considered as part of fundamental problem of its mineralogy associated with determination of the limit up to which uraninite is an analogue of synthetic UO_2 in terms of chemical and corrosion properties (Janeczek et al. 1993). A wide range of crystallochemical methods was used to study uraninite samples from the Cigar Lake deposit, Canada. X-ray diffractometry, X-ray photoelectron spectroscopy, analytical electron microscopy (AEM) methods, including SAED, have been used. Analysis of the literature and their’ own results did not allow the authors, contrary to their expectations, to obtain evidence of the natural origin of tetragonal uranium oxide U_3O_7.

At present, research is being developed related to issues of nuclear fuel disposal; while studying the matter, intense attention is paid to secondary uranium minerals. The transformation of some uranium minerals during weathering is studied (for example, Finch et al. 1992, Isobe et al. 1992, etc.); data on uranyl minerals crystallochemistry are systematize (Stohl and Smith 1981; Burns et al. 1996; Locock and Burns 2002; Burns 2003; Hottorn 2003; and others). Increased interest in hypergene uranium minerals (mainly U^{6+} minerals) is associated with consideration of the transformation process of primary uranium ore minerals as a prototype for the behavior of buried radioactive waste. In this light, study of hypergene uranium mineral-formation and the weathering crust of uranium deposits again become urgent. Environmental studies require the investigation of modern mineral-formation and, consequently, the study of dispersed mineral matter in small quantities, which is possible only when using a complex of local APEM methods.

The first experience with ASEM use in uranium minerals study was illustrative (Deliens and Piret 1977; Deliens and Piret 1985a, b) demonstrating morphology of secondary uranium phosphate formations (Al-uranyl phosphates), which were too small for light optics resolution. Since the late 1980s, researches of uranium minerals by ASEM methods have been carrying out for the ores of Northern Bohemia, the Czech Republic (Scharm 1993; Scharm et al. 1994; Scharmova et al. 1993; Scharmova and Scharm 1994). In the study of Th-U-phosphates, the largest of currently known ningyoite crystals (20–40 microns) were found and analyzed.

The author of this book has been studying the dispersed uranium minerals since the 1970s, using the optimal complex of ATEM methods (SAED, energy—and wave—dispersed spectrometry), and so developing of SAED-pictures decoding techniques in co-authorship and under the guidance of A.I. Gorshkov. The results obtained with participation of the author served as basis for hypergene uranium mineralization study and allowed to come to a whole series of *mineralogical discoveries* in co-authorship with L.N. Belova. The results of detailed uranous mineralization study by ATEM methods are:

- A new industry type of uranium mineralization has been discovered—phosphate (ningyoite) one;
- A new mineral family (supergroup)—uranous (U^{4+}) phosphates is discovered. The family consists of various mineral groups delegates and is united by close formation conditions in reductive hypergenesis region. Until our research, the single uranous phosphate, ningyoite, was known; it considered as mineralogical rarity;
- Two new mineral groups of uranium were discovered: Lermontovite group—uranous phosphates; Coconinoite group—Al-uranyl sulpho-phosphates (Doynikova and Sidorenko 2006).
  - Into mineral group of Lermontovite, whose individuality was previously questioned and was proved by ATEM methods (Melkov et al. 1983), were attributed two new U^{4+}phosphate minerals—Vyacheslavite and Urphoite—discovered by us (Belova et al. 1998).
  - Into mineral group of Coconinoite, in addition to the variety which gives the name of the group, was included Moreauite, Furongite and Xiangjiangite (Belova et al. 1993; Belova et al. 1994). The basis for combining these minerals into a single mineral group was similar optical characteristics, composition proximity, structure unit cells similarity (symmetry and parameters), and also general formation conditions: near the redox boundary in the hypergenesis region. A wide isomorphism of Al, Fe and Cr in coconinoite composition was established (Belova et al. 1993; Doynikova and Sidorenko 2006).

ATEM study of dispersed uranous phosphates led to the discovery of new mineral varieties: iron-containing version of U^4-phosphate—Fe-ningyoite and Al-version of uranyl-phosphate—Al-coconinoite. A new variety of REE-phosphate mineral—Ca-rhabdophane—was discovered (Guliy et al. 1990; Doinikova et al. 1993).

Our studies, conducted jointly with L.N. Belova, an outstanding specialist in the hypergene mineralogy of uranium, were showed the polymineral character of widespread black uranium ore mineralization (Belova et al. 1991, Belova et al. 1997; Doynikova et al. 2003). It is proved that so-called "uranium blacks" can be represented by three different mineral forms of uranium: oxide (uraninite), silicate (coffinite) and phosphate (ningyoite). They are manifested in various proportions, up to monomineral formation. Oxide form is the most common. It is shown that the traditionally used concept "uranium blacks" can characterize only morphology and dark, almost black, color of ore sample, but use of this term to characterize mineral composition will be mistakenly.

By ATEM methods in samples, visually attributed to uranium blacks, is revealed a new, previously practically unknown, uranous phosphate, ningyoite (second finding in the world) (Belova et al. 1978a). The wide distribution of this phosphate in infiltration deposits is proved. The study of mineral paragenesis and isomorphic impurities in ningyoite made it possible to make a number of assumptions about the geochemical conditions of this uranous phosphate formation (Doynikova 2003). Ningyoite ores was discovered by us as in several fields of Bulgaria (Belova et al. 1996) so in Russian deposits of Khiagda; it made possible to detect a new type of uranium ore—phosphate black one (Doynikova et al. 2014). (The results of the study of U^{4+} minerals are given below in the mineralogical section).

As a result of long-term study of uranium mineralization by ATEM methods, new information was received on the uranites (so-called "uranium micas") most widespread in nature. The processes of uranium micas substitution by iron hydroxides were studied using the SAED method; the instability of U-micas, affected by substitution or leaching processes, under electron beam is shown (Belova et al. 1981). These data are summarized in the author's PhD dissertation (Ivanova 1982). The detailed study of dark to black specimens of autunite and uranospinite not confirmed existing ideas about the nature of these minerals coloring: namely, that dark color is caused by U^{4+} oxides formation (Belova et al. 1992). The X-ray photoelectron microscopy (XPS) method has established that the reason of blackening of carnotite, which has been studied by ATEM methods, is the presence of the different vanadium forms V^{4+} and V^{5+} in its structure (Belova et al. 1983).

On example of natural brannerite study, the author demonstrated possibility of metamict minerals diffraction study (Ivanova 1982): structural characteristics of brunnerite in its natural state (without calcinations) were obtained for the first time. On the SAED basis, it is proved that brannerite is uranium titanate, rather than a complex oxide of U and Ti, as it was thought previously.

The author participated in study of Karkhu deposit, Ladoga area, which was attributed to "unconformity" type, according to localization conditions (gently sloping zones of structural and stratigraphic disagreement, shallow depths) (Velichkin et al. 2002; Velichkin et al. 2003). Here uranium ores are difficult for traditional optical diagnostics due to their micro-inhomogeneity. Such "multiphase" state of ore is due to different content of "polluting" impurities (Si, Pb, Ca, and S) and abundance of galena impregnation of submicron sizes. Only SAED method using allowed the surely diagnosing uranium formations rich in silicon ($\approx$10% Si) as poorly crystallized uraninite. Heterogeneity of this oxide uranium mineralization is represented by various spherulite-like and spherulitic formations (not less than three kinds) that have a similar composition and differ only in carbon and/or silicon presence. Enrichment of the uranium-ore phase with carbon was found in rich samples. Such spherulite morphology of uranium-oxide (with carbon) formations in combination with their composition analysis permit us to assume that there have been processes of recrystallization of ore matter, leading to purification of the formed U-pitchblende spherulites (from Pb, S in form of galena and from Si). Conclusions on the processes occurring in the ore body were made in overseeing SEM images: dissolution of quartz and pyrite observed in the same sample. Analysis of conditions allowing simultaneous dissolution of these minerals led the author to conclusion about action of heated neutral waters with free oxygen, indicating infiltrating surface waters and regional warming up of the thickness (Doynikova 2007).

The author used complex AEM methods during studying uranium ores of deposits localized in paleo-valleys of the West Siberian Plate: Malinovskoye and Khokhlovskoye. In the Malinovskoye ores, for the first time was identified predominantly silicate form of uranium (coffinite) of tiny, micron size; previously only oxide form of uranium was known here. Coffinite was diagnosed by combination of electron diffraction data (ATEM) and composition analysis (ASEM). In ores intervals selenium mineralization (ferroselite-$FeSe_2$) and intense chloritization was detected. The morphology of coffinite-pyrite inter-relations clearly fixes place of uranium (silicate) mineralization in mineral-forming process: after dissolution (oxidation) of the early pyrite and simultaneously with late generation pyrite. (Doynikova et al. 2000). Based on obtained results, already at the first stage of samples mineralogical investigation, it was concluded that this deposit ore are not subject to development by underground leaching: close association uranium ore with Fe-chlorite removes this deposit from its exploitation profitability by ISL method (Vinokurov et al. 2001).

Unlike coffinite ores of Malinovskoye, industrial uranium mineralization at Khokhlovskoye deposit is represented by uraninite. Here there is an abundance of different kinds of organic matter; uraninite phytomorphosis in carbonaceous remains is widely developed. The feature of these uranium ores is uranium oxides micro- and nano-graininess in well-permeable sands, what is of interest for ISL. (Dymkov et al. 2003). Complex study results of these poor ores both deposits published in specialized editions (in Russian).

The author's research allowed considering the issue of bitumen formation due to the polymerization of natural petroleum under radioactivity influence (Dymkov et al. 2002). Experimental confirmation of uranium mineralization effect on natural oil transformation has been obtained. Pyrobitumen films (80–120 μm), which were formed on nasturane (pitchblende) spherulites on contact with oil as result of thermobaric treatment in autoclave ($T = 3000\ ^\circ C$, $P = 195 \pm 10$ kgf/cm^2), have been visualized by ASEM methods. Bitumen formation on nasturane goes more intensively than on mourite (uranium molybdate). Bitumen crusts thickness and amount of bitumen around pitchblende grains are much higher than around the larger fragment of mourite. Actually uranium minerals radiation exposure on oil during the experiment was short, but results allow us to confidently talk about catalytic role of uranium minerals in solid bitumen formation from oil.

Radioactivity effect of natural uraninite on organic matter transformation is also traced in Early Proterozoic Au-U ores of Witwatersrand Basin, South Africa (samples of Dr. Yu.G. Safonov). ASEM using allowed obtaining additional facts confirming conclusions on primary sedimentary nature of gold-bearing "reefs". A sample of «bonanza» ores (carbonaceous thin beds up to 5 cm) was studied (Safonov et al. 2000). Here unchanged uraninite cubic crystals, perfect form often, are embedded in bitumen (kerogen) formations. Peculiar bituminous columnar forms cross the layers (perpendicular to the sole) and create coarse (~ 200 μm) striations' pattern of black color. Less often bitumen envelops nasturane grains as "shirt", filling and healing cracks in nasturane crystals. These formations morphology is surprisingly reminiscent an algal origin fossil—Spongiostromata group, the cyanobacteria (cainophyta) department (Mein 1987). This columnar carbon formations morphology is very close to stromatolites, beside internal stratified structure, which was not observed. It is assumed that they were originally mini-stromatolites (oral message PhD M.E. Raaben, GIN RAS). Typical for stromatolites sedimentation conditions (shallow sea lagoon facies) suggest that these modern bitumen formations initially existed in the Early Proterozoic as algae or cyanobacteria (stromatolites). Such interpretation is also based on conclusion of plant organics matter conservation as kerogen (Dr. M.A. Fedonkin, PIN RAS, oral communication). Uraninite crystals presence embedded in the bituminous columnar "interlayers" is well explained by mechanical migration of uraninite and its micro crystals accumulation in plant benthonic mass. In historical geochemistry of uranium (Perelman 1989) it is well known accumulation of uraninite (initially hypogenous) as placers during reductive Archean—Lower-Proterozoic stage in the Earth development, when there was no oxygen in atmosphere. Specific schistosity/lamination of carbonaceous substance (kerogen) observed in a samples of Au-U-ores of Witwatersrand, can be explained (at this stage study), by radioactive irradiation action (by spontaneous decay uranium products) on the buried organic material.

The experience of uranium mineralization studying by ATEM and ASEM methods is allowed the author to generalize some methodological aspects important for mineralogical observations. Statistics of uranium ores study, different in the ore component content, showed that the regularities of uranium mineralization distribution in them are the same. To research mineral formation processes, rich ore samples are preferred, in which precipitation mechanisms are the same as in poor samples.

The using analytical electron microscopy as mineralogy method allowed the author not only make a number of mineralogical discoveries, but also served as the main source of crystallochemical data necessary to refine development regularities of redox uranium mineral zonality in hypergenesis region, namely in lowest horizons and in secondary enrichment sites (Belova 2000; Doynikova 2003; Doynikova et al. 2003). The results of TEM-study dispersed uranium minerals from many different samples were collected and analyzed in the author's Habilitation Thesis (Doynikova 2005) and then published in book (Doynikova 2012). These results of study uranium mineral-formation under reducing conditions supplemented by consideration of a new aspect of hypergene uranium ores appearance, biogenic one, are presented in the following chapters of the book.

References

Andrews K, Dyson D, Clown S (1971) Electron diffraction patterns and their interpretation. M.: Mir. [In Russian]

Avilov AS (2003a) Electron diffraction structure analysis // electron crystallography on solid state inorganic materials and nanostructures. Moscow (EURO-)Summer School on electron crystallography. Shubnikov Institute of Crystallography, Moscow, pp 1–10

Avilov AS (2003b) The quantitative analysis of electrostatic potential by electron diffraction structure analysis. Ibid, Moscow, pp 81–95

Belova LN (2000) Formation conditions of oxidation zones of uranium deposits and uranium mineral accumulations in the hypergenesis zone. Geol Ore Deposits 42(2):113–121. [in Russian]

Belova LN, Doynikova OA, Gorshkov AI, Sivtsov AV (1996) New variety in the mineral group of coconinoite: Al-coconinoite. Proceedings annual VMO. Session. M. P. 76a [in Russian]

Belova LN, Gorshkov AI, Doinikova OA, Sivtsov AV, Dikov YP, Lyubomilova GV (1992) New data on U (IV)–containing micas. Doklady RAS T 325(1):139–141. [In Russian]

Belova LN, Gorshkov AI, Doynikova OA, Sivtsov AV (1991) Uranium black. In: Book: "development of V.I. Vernadsky's ideas in geological sciences". Nauka, Moscow, pp 165–172. [in Russian]

Belova LN, Gorshkov AI, Doynikova OA, Sivtsov AV (1994) A new mineral group: Coconinoite and other (Al, Fe)-uranyl-sulpho-phosphates. Intern. Miner. Assoc. 16-th General Meeting, Abstr. Pisa. Italy. P. 36

Belova LN, Gorshkov AI, Doynikova OA, Sivtsov AV (1997) Contribution of electron microscopy to the development of uranium mineralogy. Intern. Symposium on 100th anniver. IGEM RAS, A.G. Betekhtin Moscow, pp 213–214

Belova LN, Gorshkov AI, Doynikova OA, Sivtsov AV (1998) A new group of minerals–phosphates U (IV). DAN USSR 358(2):215–217. [in Russian]

Belova LN, Gorshkov AI, Doynikova OA, Sivtsov AV, Mokhov AV, Trubkin NV (1993) New data about coconinoite. Doklady AN 329(6):772–775. [in Russian]

Belova LN, Gorshkov AI, Doynikova OA, Sivtsov AV, Zelenova O, Filippov IM, Drahomanov LV (1982) Ningyoite in Bulgaria. Crystal chemistry of minerals. International Mineralogical Association (IMA): Proceed. 13 General Meeting, Varna Ed.: Bulgarian Academy of Sciences. 1986. Vol. 1. P.773–779

Belova LN, Gorshkov AI, Ivanova OA (1976) Substitution of uranium micas with iron hydroxides. Izvestiya AN SSSR ser geol 12:151–157. [in Russian]

Belova LN, Gorshkov AI, Ivanova OA (1978a) The first find of ningyoite in the USSR. Doklady AN USSR 238(1):215–216. [in Russian]

Belova LN, Gorshkov AI, Ivanova OA (1978b) New data on ningyoite: Fe-containing ningyoite. Dokl AN USSR 243(4):1022–1023. [in Russian]

Belova LN, Gorshkov AI, Ivanova OA, Dikov YP, Sivtsov AV, Ryzhov BI (1983) On V^{4+}–containing carnotite. Izvestiya AN SSSR, ser geol 11:124–128. [in Russian]

Belova LN, Gorshkov AI, Ivanova OA, Sivtsov AV (1981) About uranium micas (uranites) containing tetravalent uranium. Izvestiya AN SSSR ser geol 1:139–142. [in Russian]

Belova LN, Gorshkov AI, Ivanova OA, Sivtsov AV (1985) Ningyoite in the light of new experimental data. Izvestiya AN SSSR Ser Geol 3:101–109. [in Russian]

Bokiy GB (1971) Kristallokhimiya. MSU, Moscow, p 319. [in Russian]

Bokiy GB, Poraj-Koshits MA (1964) X-ray structural analysis. Vol.1. Ed. 2. Moscow, MSU, p 489. [in Russian]

Burns PC (2003) Recent advances in the uranium minerals structures study and uranium crystal chemistry. Zapiski VMO. Part 132. pp. 90–114. [in Russian]

Burns PC, Miller ML, Ewing RC (1996) U^{+6} minerals and inorganic phases: a comparison and hierarchy of crystal structures. Canad. Fortschr Mineral V(34):845–880

Chukhrov FV, Gorshkov AI, Drits VA (1989) Hypergenic manganese oxides. Nauka, Moscow. [in Russian]

Cliff G, Lorimer GW (1975) The quantitative analysis of thin specimen. J Microsc 103(2):203–207

Cowley JM (1967) Crystal structure determination by electron diffraction. Prog Mater Sci, Pergamon Press, Oxford 13(6):267–321

Deliens M., Piret P. Les phosphates d'uranyle et d'aluminium de Kobokobo. I. Données preliminaries. Bull Soc Belg Geol. 1977. 86. ¾. 183–190

Deliens M, Piret P (1985a) Les phosphates d'uranyle et d'aluminium de Kobokobo. VIII. La Furongite. Ann Soc Geol Belg 108:365––368

Deliens M, Piret P (1985b) Les phosphates d'uranyle et d'aluminium de Kobokobo. VII. La Moreauite, $Al_3UO_2(PO_4)_3(OH)_2 \cdot 13H_2O$, nouveau mineral. Bull Mineral 108(1):9–13

Dorset DL (1995) Structural electron crystallography. Plenum Press, New York. 452 p

Doynikova OA (2003) Genetic Cristal chemistry of the mineral components of uranium blacks. Geochem Int 41(12):1214–1220

Doynikova OA (2005) Dispersed uranium minerals of hypergenesis reducing zone: mineralogy and crystallochemistry. Habilitation Thesis. M.: IGEM RAS. 277 p

Doynikova OA (2007) Colloform nature of uranium ores in deposit Karkhu as a genetic marker. New ideas in the earth sciences, vol 3. VNIIgeosistem, Moscow, pp 107–110. [In Russian]

Doynikova OA (2012) Uranium mineralogy in Hypergenesis reduction zone (according to electron microscopy data). Fizmatlit, Moscow, p 216. [in Russian]

Doynikova OA, Belova LN, Gorshkov AI, Sivtsov AV (2003) Uranium blacks: questions of genesis and mineral composition. Geology of Ore Deposit 45(6):514–530. [In Russian]

Doynikova OA, Gorshkov AI, Belova LN, and others (1993) Questions on systematics of phosphates rhabdophane group. Zapiski VMO 3:79–88. [In Russian]

Doynikova OA, Sidorenko GA (2006) Coconinoite and related alumosilicates. New Data on Minerals 41:49–61

Doynikova OA, Sivtsov AV, Vinokurov SF (2000) New in the Mineralogy of uranium ores on Malinovskoe Deposit. Proc. Session VMO "Mineralogy of Russia". SPb. Abstr. pp. 109–110 [In Russian]

Doynikova OA, Tarasov NN, Mokhov AV (2014) A New Phosphatic type of uranium deposits in Russia. Dokl Earth Sci 457(Part 2):910–914. https://doi.org/10.1134/S1028334X14080030

Drits VA (1981) Structural study of minerals by microdiffraction and high-resolution electron microscopy. Nauka, Moscow, p 240. (in Russian)

Drits VA (1987) Electron diffraction and high resolution electron microscopy of mineral structures. Springer, Heidelberg., N.-Y, p 304

Drits VA, Kameneva MY, Sakharov BA, Dainyak LG, Tsipursky SI, Smolyar BB, Bukin FS, Salyn AI (1993) Problems of determining the real structure of glauconites and related fine-dispersed phyllosilicates. Novosibirsk, VO Nauka, p 200. [In Russian]

Drits VA, Lanson B, Bougerol-Chaillout C, Gorshkov AI, Manceau A (2002) Structure of heavy-metal sorbed birnessite: part 2. Results from electron diffraction. Am Mineral 87:1646–1661

Drits VA, Silvester E, Gorshkov AI, Manceau A (1997) Structure of monoclinic Na-birnessite and hexagonal birnessite: I. results from X-ray diffraction and selected-area electron diffraction. Am Mineral 82:946–961

Dubinchuk VT, Belyaevskaya NG (1975) Electron microscopic study of brannerite. Mineralogical digest of Lviv Geol Soc 4(29):61–64

Dubinchuk VT, Kochenov AV, Ruzhitsky VV, Meshchankina VI (1990) Forms of precipitate uranium mineralization of exogenous epigenetic ore-formation in sedimentary rocks according to electron microscopic studies. Lithol Miner Resour 3:65–72. [In Russian]

Dubinchuk VT, Naumova IS, Kravtsova IY, Sidorenko GA (1981) Refinement of coffinite crystal structure. Mineral J 4:81–85. [In Russian]

Dymkov YM, Doynikova OA, Volkov NI (2003) Found of Fe-Zr-Ti-SP-gel in exogenous-epigenetic uranium deposit Khokhlovskoye (southern trans-Urals). Geochemistry 11:62–67. [In Russian]

Dymkov YM, Kuntz AF, Doinikova OA (2002) Solid bitumen formation during interaction of nasturan and oil at 300 ° C. Dokl RAS 387(1):90–94. [In Russian]

Finch RJ, Miller ML, Ewing RC (1992) Weathering of natural uranyl oxide hydrates: schoepite polytypes and dehydration effects. Radiochemistry Acta 58(59):433––443

Gorshkov AI (1970) The use of electron microdiffraction to obtain basal reflexes from lamellar layered silicates. Izvestiya AN SSSR Ser geol 3:133–136. [in Russian]
Gouldstein J, Yakovits HM (1978) Practical raster electron microscopy. Mir, Moscow, p 362. [in Russian]
Gritsaenko GS, Rudnitskaya ES, Gorshkov AI (1961) Electron microscopy of minerals. Publishing house USSR Academy of Sciences, Moscow, p 132. [in Russian]
Gritsaenko GS, Zvyagin BB, Boyarskaya RV et al (1969) Methods of electron microscopy of minerals. Nauka, Moscow, p 311. [in Russian]
Guliy VN, Doynikova OA, Furuta T, Sivtsov AV (1990) Mineral inclusions in apatite of metamorphic rocks and ores of the Aldan shield. Mineralogical digest of Lviv state Univ Lviv 2 (44):69–73. [in Russian]
Hirsch P, Howie A, Nicholson R, Pashley D (1968) Electron microscopy of thin crystals, Moscow, Mir, p 574. [in Russian]
Hottorn FK (2003) Valences balance: a new approach to structure, chemical composition and paragenesis of minerals-oxysalts with hydroxyl groups and water molecules. Geol Ore Deposits 45(2):100–133. [in Russian]
Huang DX, Liu W, Gu YX et al (1996) A method of electron diffraction intensity correction in combination with high resolution electron microscopy. Acta Cryst A52:152–157
Isobe H, Murakami T, Ewing RC (1992) Alteration of uranium minerals in the Koongarra deposit, Australia: Unweathered zone. J Nucl Mater 190:174–187
Ivanova OA (1982) Electron microscopic study of fine-dispersed uranium minerals. PhD dissertation. IGEM RAS, Moscow, p 172. [in Russian]
Ivanova OA, Melnikov IV, Gorshkov AI, Vampilov MV (1982) Brannerite in the light of data of analytical electron microscopy. Izvestiya AN SSSR Ser Geol 2:63–71. [in Russian]
Janeczek J, Ewing RC, Thomas LE (1993) Oxidation of uraninite: does tetragonal U_3O_7 occur in nature? J Nucl Mater 207:177–191
Kerr PF (1951) Natural black uranium powder. Science 114:91–95
Kolb U et al (2007) Towards automated diffraction tomography: part I--data acquisition. Ultramicroscopy 107:507–513
Locock AJ, Burns PC (2002) Crystal structures of three framework alkali metal uranyl phosphate hydrates. J Solid State Chem 167(226):52–61
Mein SV (1987) The basics of paleobotany. Nedra, Moscow. [in Russian]
Melkov VG, Belova LN, Gorshkov AI, Ivanova OA, Sivtsov AV (1983) New data on lermontovite. Mineral J 5(1):82–86. [in Russian]
Moeck P, Rouvimov S, Häusler I, Neumann W Nicolopoulos S (2011) Precession electron diffraction & automated crystallite orientation/phase mapping in a transmission electron microscope. (2011) 11th IEEE International Conference on Nanotechnology Portland Marriott, August 15–18, Portland, Oregon, USA, pp. 754–759
Mokhov AV (1986) Method for analysis of mineral microparticles in a transmission electron microscope. Izv. AN SSSR, ser. geol., No. 4, pp. 99–104. [(in Russian]
Mokhov AV (1987) Diagnostics of finely dispersed minerals by analytical electron microscopy. Ph. D dissertation. IGEM RAN. Moscow.183 p. [in Russian]
Mokhov AV, Tsepin AI (1979) Technique for Processing X-Ray semiconductor Spectra in Transmission Electron Microscopy. VII Russ. Conf. on local x-ray spectral studies. Chernogolovka. Abstr. p. 56. [in Russian]
Mokhov AV, Tsepin AI (1980) The technique of semi-quantitative analysis of minerals in a transmission electron microscope equipped with a semiconductor detector. Izvestiya AN SSSR ser fiz 44(10):2183–2185. [in Russian]
Muto T, Meyrowitz R, Pommer AM, Murano T (1959) Ningyoite–a new uranous phosphate mineral from Japan. Am Mineral 44(5–6):633–639
Nekrasov IY, Doinikova OA, Gorshkov AI, Sivtsov AV (1987) The first find of Kainosite in the USSR. Docl AN SSSR 294(4):948–952. [in Russian]
Perelman AI (1989) Geochemistry. Moscow, Higher school, p 574. [in Russian]

Pinsker ZG, Goodman P (1981) Fifty years of electron diffraction. D. Reidel. Pub. Co, Dordrecht, pp 155–163

Pyatenko YA, Voronkov AA, Pudovkina ZV (1976) Mineralogical crystal chemistry of titanium. Nauka, Moscow, pp 37–67. [in Russian]

Reed S (1979) Electron-probe microanalysis. Mir, Moscow, p 216. [in Russian]

Ruh R, Wadsley AD (1966) The crystal structure of $ThTi_2O_6$. Acta Cristallogr 21:483–490

Safonov YuG, Bershov LV, Bogatyrev BA, Gorshkov AI, Doynikova OA, Zhukov VV (2000) Main signs of the primary sedimentary nature of early Proterozoic gold-uranium ores of Witwatesrand basin (South Africa). "Natural and technogenic placers and crustal weathering deposits at the turn of the Millennium". XII Intern. Meeting on the Geology of placers and weathering crusts. Abstracts. M. pp. 325–328. [in Russian]

Scharm B (1993) Některé vzácné minerály provázející uranové zrudnění v severočeské křídě. Bull mineral-petrogr Odd Nár Muz Praha:45–48

Scharm B, Scharmova M, Kundrát M (1994) Crandalite group minerals in the uranium ore district of northen Bohemia (Czech republic). Věst Čes Geol Ust Praha 69(1):79–85

Scharmova M, Scharm B (1994) Rhabdophan group minerals in the uranium ore district of northen Bohemia (Czech Republic). J Czech Geol Soc 39(4):267–280

Scharmova M, Scharm B, Rutšek J (1993) Rabdofan-(Nd), churchit-(Y) a langizit na ložisku Stráž v severočeské křídě. Sbor.V. Mineralogický cyklický seminář. Ústí na Labem. Audit VT. S. 9799

Schimmel G (1972) Technique of electron microscopy. M. Mir, 1972 [in Russian]

Shau Y-H, Peacor DR (1992) Phyllosilicates in hydrothermally altered basalts from DSDP hole 504B, leg 83–a TEM and AEM study. Contribution to Mineralogy and Petrology 112:119–133

Sidorenko GA (1978) Crystal chemistry of the uranium minerals. Atomizdat, Moscow, p 216. [in Russian]

Sidorenko GA, Gorobets BS, Dubinchuk VT (1986) Modern methods of mineralogical analysis of uranium ores. Energoatomizdat, Moscow, p 184. [in Russian]

Stohl FV, Smith DK (1981) The crystal chemistry of the uranyl silicate minerals. Am Mineral 66:610–625

Tsirelson VG, Avilov AS, Abramov YA, Belokoneva EL et al (1998) X-ray and electron diffraction study of MgO. Acta Cryst B54:8–17

Utevsky LM (1973) Diffraction electron microscopy in metal science. Metallurgy, Moscow, p 583. [in Russian]

Vainshtein BK (1956) Structural electron diffraction. Akad. Nauk SSSR, Moscow, p 314. [in Russian]

Vainshtein BK (1964) Structure Analysis by Electron Diffraction (Pergamon, 1964)

Vainshtein BK, Zvyagin BB, Avilov AS (1992) Electron diffraction structure analysis. In: Cowley JM (ed) Electron diffraction techniques, vol 1. Oxford Univ. Press, Oxford, pp 216–312

Vainstein BK (1979) Modern crystallography (in 4 volumes). In: Crystal symmetry. Methods of structural crystallography, vol 1. Nauka, Moscow, p 384. [in Russian]

Vainstein BK, Fridkin VM, Indenbom VL (1979) Modern crystallography, vol V. 2. Nauka, Moscow, p 367. [in Russian]

Velichkin VI, Tarasov NN, Andreeva OV, Golovin VA, Golubev VN, Doynikova OA, Kiseleva GD, Krylova TL (2002) Geological, mineral-petrography and physical-chemical features of the polycomponent uranium Deposit Karkhu (Northern Ladoga region). "Geology, Genesis and issues of development of complex minerals of precious metals". All-Russia Symposium IGEM RAN. Abstr. Moscow: OOO "Svyaz' –Print". P.108–109. [in Russian]

Velichkin VI, Tarasov NN, Andreeva OV, Kiseleva GD, Krylova TL, Golubev VN, Doinikova OA, Golovin VA (2003) Geology and formation conditions of Karkhu deposit—the first uranium "unconformity" deposit in Russia. Proc. AN "URANIUM GEOCHEMISTRY 2003" Nancy, France (Proc. Intern. conf.): Geochimie de l'uranium 2003–Nancy. P. 371–374

Vinokurov SF, Doinikova OA, Krylova TL et al (2001) Lithological, geochemical and mineralogical features of the Malinovskoye uranium deposit (Russia). Geol Ore Deposits 43(5):414–429. [in Russian]

Weirich TE (2003) Electron diffraction structure analysis: structural research with low-quality diffraction data. In: Electron crystallography on solid state inorganic materials and nanostructures. (EURO-SUMMER SCHOOL) Moscow Summer School on electron crystallography. Shubnikov Institute of Crystallography, Moscow, pp 20–43

Weirich TE, Hovmöller S, Kalpen H et al (1998) Electron diffraction versus X-ray diffraction–a comparative study of the Ta_2P structure. Crystallography Reports 43:956–967

Weirich TE et al (2000) Structures of nanometer-size crystals determined from selected-area electron diffraction data. Acta Cryst A56:29–35

Weirich TE et al (2002) Structure of nanocrystalline anatase solved and refined from electron powder data. Acta Cryst A58:308–315

Zhukhlistov AP, Avilov AS, Ferraris D, Zvyagin BB, Plotnikov VP (1997) Evidence of a three-position disordered distribution of hydrogen in the structure of brucite mg $(OH)_2$ according to electron diffractometry. Crystallography 42(5):841–845. [in Russian]

Zvyagin BB (1964) Electronography and structural crystallography of clay minerals. M: Science:282. [in Russian]

Chapter 2
U-Ore Mineralization of Hypergenesis Reduction Zone: Uranium "Blacks"

Disperse uranium mineralization in form of loose powder is manifested in almost all types of uranium deposits, especially in hypergenesis zone. In uranium mineralogy, dispersed friable sooty radioactive mineral-formation is called "black" (niello); it may of black or brownish-black color. In Russian geological terminology such black U-mineralization is called "cherny" or "chernevye" U ores. In foreign literature, such ores are called «sooty uraninite» despite of their polymineral nature.

At hydrothermal deposits, uranium blacks are formed by weathering of sulfide-containing uranium ores in lowest zone of hypergenic processes. Ores of hydrogenic (exogenous-epigenetic) uranium deposits in sedimentary cover are everywhere composed of uranium blacks. In deposits of sedimentary cover, uranium black is of greatest industrial interest, as it is economically feasible to develop such deposits using the ISL (in situ leaching) method.

Black mineralization is often the main and even the only one component of commercial uranium deposits. Among hydrogenic uranium deposits sandstone-type ones dominate in world uranium mining. They are most cost-effective for in situ recovery. In 2015, more than half world uranium production (about 55%) was extracted from deposits associated with sandstones (Uranium 2016). These exogenous ores are formed as a result of sediments stratal−/ground—oxidation by infiltration of oxygen-rich uranium-bearing groundwater. In accordance with IAEA classification (adopted in 2013), among sandstone associated uranium deposits, three subtypes are most widely represented: tabular—Czech Republic (Strazh), Bulgaria (Marishsky ore field); roll-front—Kazakhstan (ChuSarysu type); basal channel—Japan (Ningyo-Togo, Tono), Canada (Blizzard, British Columbia), Argentina (Mendoza), Russia (Khiagda ore field, Trans-Baikal region; Dalmatovo, Trans-Ural region) etc.

This black uranium-ore mineralization of sandstone type deposits is difficult to diagnose by traditional mineralogy methods. However, undoubted interest in their mineral composition is caused by their wide spread and manifestation in ores of various genetic types.

O. A. Doynikova, *Uranous Mineralogy of Hypergene Reduction Region*,
Springer Mineralogy, https://doi.org/10.1007/978-3-030-67183-9_2

Our long-term experience in such loose black U-mineralization study has proved high efficiency of analytical electron microscopy (AEM) in analysis of mineral composition of these ores. Results of these mineralogical AEM studies form a core of this and next chapter.

2.1 Regularities of Black's Uranium Mineralization Manifestation in Hypergenesis Region

On the basis of results long-term field observations and mineralogical studies, with literature data use, Russian mineralogist L.N. Belova (1923–1998) has created a Systematics of exogenous ore concentration of uranium (Belova 2000), generalizing the mineralogy of oxidized (uranyl) forms of uranium. /The article was prepared for publication by the author together with A.V. Timofeev on materials L.N. Belova/ Systematics included the classification of oxidation zones of hydrothermal uranium deposits (Belova 1975), as its integral part. The data on the regularities of formation of uranium ore accumulations in the hypergenesis region (for oxidation zones), combined by L.N. Belova in the table of typical mineral zones in the weathering crusts of deposits, were specified by adding the reducing (black/blackish) mineral zone (Table 2.1) (Belova and Doinikova 2003a, b). Later, according to L.N. Belova's materials (Belova 2000), this table was extended by include examples of deposits and ore occurrences, where the most obvious and wide represented in nature corresponding types of oxidation/reducing zones is given (right column of Table 2.1) (Doynikova 2005).

In the paper (Belova 2000) conclusions about process development peculiarities, that forming mineral zonality on uranium deposits in hypergenesis zone, are made from the generalized consideration of mineral formation regularities. Given by L.N. Belova in text form, the staged processes of uranium mineral formation, was presented in schematic form for clarity. The idea of schematization belongs to A.V. Timofeev. The scheme of weathering stages with the addition of reducing (blacks') stage is given in (Doynikova et al. 2003). This scheme published here was later supplemented by another type of ore—infiltration one (Doynikova 2005, 2012a, b). Thus, here on scheme (Table 2.2) all types of uranium *ore* concentration, manifested in hypergenesis region, which united by general laws of mineral formation, are presented.

To characterize location of reduction uranium mineralization (blacks) in hypergenesis region, the conditions of hypergene uranium mineral-formation are briefly considers below.

Table 2.1 Types of exogenous concentrations of uranium-ore mineralization (according to L.N. Belova 2000), position of black mineralization in natural sequence of mineral zones

Deposits types (by composition of primary ores)	Ore concentration types	Types of oxidation/**reduction** zones and ore occurrences	Deposits and ore occurrences where the corresponding zone types are most clearly represented
Uranium: Uranium proper (sulfide-free)	I "Hats on the head"	Hydroxide-silicate	Streltsovskoe (Transbaikalia); Yellow River (Ukraine); Panfilovskoye, Akkan-Burluk (Kazakhstan).
		Silicate	Beshtau (N. Caucasus); Chauli (Cent. Asia); Swallow (Far East).
Sulphide-Uranium *(U-ore >> sulfides)*		Silicate	Kara-Kiya (Kazakhstan); Kyrgyzstan; Northern Tien Shan.
		Uranite-silicate	Jerkamar (Central Asia); Transbaikalia.
		Uranite	Taboshar (Tajikistan); Adrasman, Charcazar (Central Asia).
Uranium-Sulfide *(sulfides >> U-ore)* *(U << Mo)*		Limonite	Beshtau (North Caucasus); Adrasman (Central Asia).
		Uranium -molybdate	Tulukuy, Luchistoye (Transbaikalia); Kyzyl-Sai (Kazakhstan).
		Blacks'	***Taboshar (Tajikistan); Kosachinoye (North Kazakhstan)***
(U << Mo) **Sulfide** (uranium-free, sharply subordinated U) + *MoS_2, PbS, selenides, tellurides*	II "Hats from other's head"	Uranium-molybdate	Katta-Sai (Central Asia).
		U-Pb-phosphate	Western Lachot (France); Jerkamar (Central Asia).
		U-Pb-arsenate	Schwarzwald (Germany).
		Uranium-selenite	Musanoy (Katanga, Zaire), [Cu-Ni-mine near U-deposit Kolwezi].
		Uranium-tellurite	Moctesuma (Mexico).
"Rootless" unconnected to uranium deposits	III "Hats without head"	Uranite	Sila Plateau (Italy); Mountain Spokane (USA); Fojtow (Czech Republic); Antraigues (Aveyron, France).
		Silicate	Mountain (Center. Transbaikalia); Beshtau (North Caucasus); Belmeken, White Iskar (Bulgaria).
Infiltration		***Blacks'***	***Chu-Sarysuy Ore Field, Kyzyl-Kum Ore Field***

2.1.1 Formation of Uranium-Ore Deposits in Hypergenesis Region (after L.N. Belova 2000)

Ore Mineral Zonality

Three *types* of uranium ore accumulations in the hypergenesis region have been identified by L.N. Belova (Table 2.1). The *first type* is represented by uranium deposit oxidation zones. The *second type* of accumulation includes oxidation zones of sulphide deposits, which serve as a kind of filter that precipitates uranium from seeping waters. The source of uranium in most cases is the adjacent crystalline massifs (mainly granite). The *third type* of uranium-ore accumulations is represented by the mineralization of oxidized uranium in crystalline massifs; it is characterized by the absence of primary ores at the depth and, for this reason, is conditionally called "rootless". Using the terminology of S.S. Smirnov (1955), a well-known Russian researcher of oxidation zones, L.N. Belova figuratively named these three types of ore uranium concentrations: "hats on the head", "hats from other's head" and "hats without a head", respectively. The table of types of exogenic uranium-ore concentration (Table 2.1) contains examples of deposits and ore occurrences where the corresponding types of oxidation or reduction zones are most clearly shown.

Black sooty mineralization, which is the subject of this study, is manifested in the first and third type of concentration.

Formation of blacks in the *first type* of concentration should be clarified, because the laws of oxidation zones formation here (within this type of concentration) significantly differ for the proper uranium and sulfide-uranium deposits.

In oxidation zones of proper-uranium deposits (unsulphide and low-sulphide), there is no secondary enrichment zone, as a rule (Fig. 2.1). Here, secondary uranium mineralization occurs directly at the place of the primary one, inside the primary ore contour: primary minerals are successively replaced by various uranium hydroxides and silicates. The oxidized parts retain the morphology of the primary ores, which indicates that there is minor uranium removal outside the ore bodies. As a consequence, it is possible to quantify the scale of primary mineralization by secondary oxidation zones.

Sulfide-uranium deposits (Fig. 2.2) are characterized by a clear *vertical* zoning and presence of the reducing zone of secondary enrichment (also known as the cementation zone). Mineral zoning, observed as the oxidation zone develops into depth: from the association of uranium hydroxides and silicates through U-micas to the secondary enrichment zone.

At weathering of sulphide-uranium ores, dependence of mineral composition of oxidation zones and degree of manifestation (thickness) of cementing zone on degree of sulphidity of ores is accurately traced. The formation of secondary enrichment zone is most typical for ores with increased sulphidity, i.e. for oxidation zones of uranium-sulfide deposits. These deposits are characterized by intensive uranium removal and dilution of ore bodies; the primary ore contours are eroded. This indicates an intensive removal of uranium outside the oxidation zone, and regenerated black mineralization is formed in the cementing zone. These signs

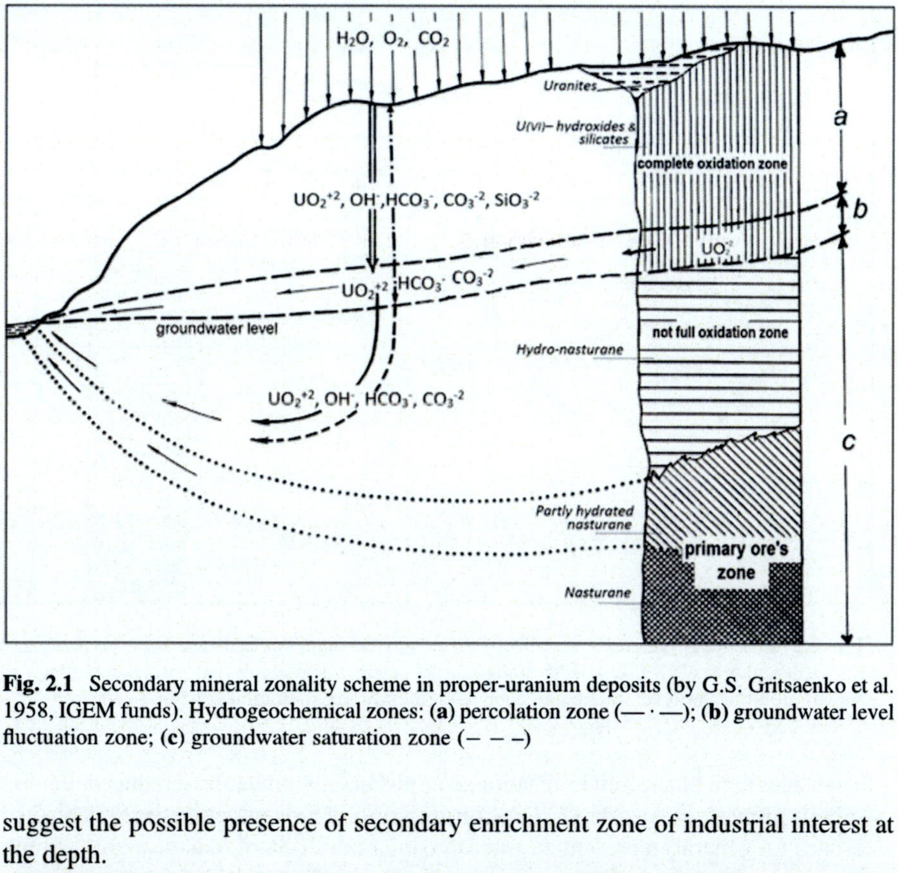

Fig. 2.1 Secondary mineral zonality scheme in proper-uranium deposits (by G.S. Gritsaenko et al. 1958, IGEM funds). Hydrogeochemical zones: (**a**) percolation zone (— · —); (**b**) groundwater level fluctuation zone; (**c**) groundwater saturation zone (– – –)

suggest the possible presence of secondary enrichment zone of industrial interest at the depth.

For deposits of the *second type* of concentration, evolution of hypergenic uranium mineral zoning is similar to sulfide-uranium ore weathering scheme. Consequently, mineralization of reduced uranium can also be expected here; however, these aggregations are unlikely to reach industrial scale, as the primary ores themselves (usually sulphide ores) are not uranium ores.

The *third type* of ore concentration in hypergenesis region, "rootless", is represented by mineralization of the oxidation zone (after Belova 2000). Typical representatives of this third type of deposits are accumulation of U-micas (autunite, meta-autunite, less often uranocyrcite) and uranyl silicates in the hypergenesis zone, reaching industrial concentrations (Table 2.1). They are usually associated with the close nearness of granite massifs or uranium deposits with a characteristic intensive uranium removal.

Black mineralization of reduced uranium is formed both in the lower horizons when U-sulfide deposits weathering and in infiltration deposits. Ores of infiltration U-deposits are conditionally referred by the author to the *third type* of concentration (Table 2.1), because they can be considered as "rootless". It should be noted that Larisa N. Belova placed in this subdivision ore accumulations composed only minerals of oxidized uranium (uranyl). As some similarity, it is probably legitimate

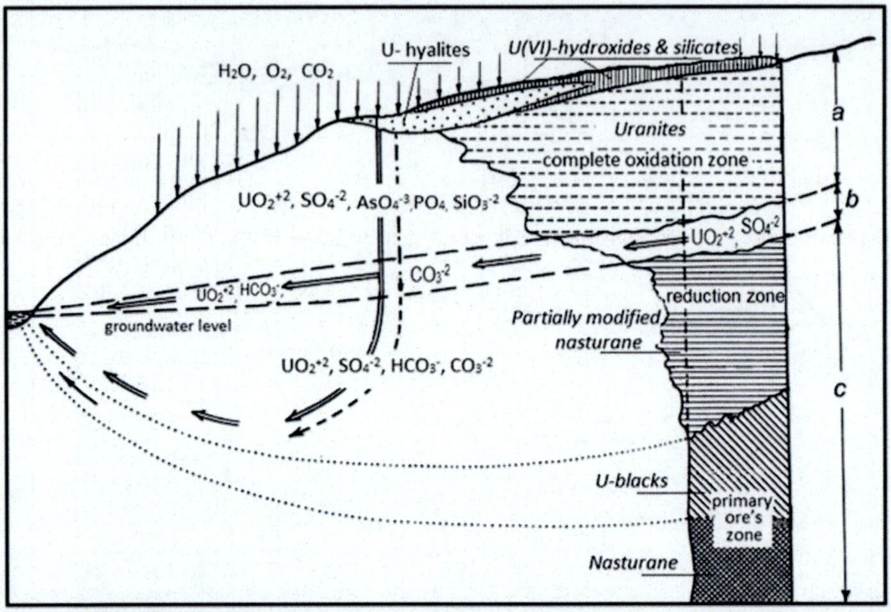

Fig. 2.2 Secondary mineral zonality scheme in uranium-sulfide deposits (according to G.S. Gritsaenko et al., 1958, IGEM funds). Hydrogeochemical zones: (**a**) percolation zone (— · —); (**b**) groundwater level fluctuation zone; c—groundwater saturation zone (– – –)

to consider here blacks mineralization, composed only minerals of reduced uranium, without primary ores at depth. Ores of infiltration (exogene-epigenetic) deposits in essence (in general) represent an extensive mineralization of reductive zone. Similar to the oxidation of sulfide-uranium ores, the scale of mineralization depends on the presence of reducing factors, which are sulfides and organic matter in ore-bearing rocks (Kislyakov and Shchetochkin 2000).

The classification of infiltration deposits ores as the third type of concentration is to some degree conditional and is necessary only to identify common, most typical futures, because there is a variety of ore concentrations (manifestations) in nature, individual for each deposit. Thus, for example, the vertical zoning (from uranium silicates through U-micas to the secondary enrichment zone) typical for the weathering of sulfide-uranium ores is traced in the horizontal zoning of uranium-ore mineralization at Sugraly infiltration deposit, Kyzyl Kum ore province (Kislyakov and Shchetochkin 2000). This fact confirms the general development regularities of uranium mineral hypergene zonality under oxygen-containing surface waters influence.

2.1.2 *Processes of Hypergenic Zonality Formation by Uranium Minerals (after L.N. Belova 2000)*

At weathering of U-ores, under the influence of oxygenated surface waters, there is a pseudomorphic substitution of primary ores with secondary uranium minerals to form redox zonality, reviewed in the table (Table 2.1).

As was determined by L.N. Belova, the main factor determining the development of the oxidation zones of uranium deposits and the manifestation of secondary zonality is the fact that uranium minerals turn out to be less stable than sulfides at weathering. The scheme (Table 2.2) shows from top to bottom the sequence of secondary uranium mineralization, which reflects changes in the mineralization medium along the direction of oxygen water infiltration (filtration).

Stadiality in the development of the mineral-forming process is associated with a continuous change in acid-alkali (pH) and redox (Eh) conditions. The degree of sulfide content in primary ore determines different directions of stadiality development (indicated by arrows) and different variants of appropriate mineral zonality.

When considering the process of interaction of sulfide-uranium ores with natural solutions, four successive stages have been identified: three oxidative and reduction stages (Belova and Doinikova 2003a). Description of oxidative stages is given further on L.N. Belova (1975, 2000); their consideration together with reductive stage is necessary to holistic view the processes of secondary uranium mineral-formation in the hypergenesis region.

The first stage—oxidation of only primary uranium minerals—occurs in a neutral or alkaline environment. They are hydrated and replaced with hydroxides and, to a lesser extent, with silicates of uranyl. Gradual change of primary uraninite (nasturane) consists in increasing degree of hydration and gradual substitution of uranium oxide by hydroxides (clarkeite, curite, fourmarierite and schoepite, minerals from groups of becquerelite, wölsendorfite and urgite). Their cationic composition (in addition to uranium) is determined by the composition of the host rocks and vein minerals. The more alkaline conditions during the weathering in the deposit are the more uranium hydroxides are formed in the first stage. The formation of specific hydroxides, or rather their cationic composition (in addition to uranium), is determined by the composition of solutions, which, in turn, is determined by the composition of the host rocks and vein minerals. At the same time, primary ores are being replaced by uranium silicates (uranophane and beta-uranophane, soddite, sclodovskite, kazolite) since neutral and slightly alkaline natural waters always contain small amounts of silica. So-called "gummite" edges and rims are formed. Since the solubility of SiO_2 reaches its maximum in a highly acidic and highly alkaline environment, the number of uranium silicates increases with the alkalinity decrease (i.e., with acidity increase during sulfide oxidation).

The formation of gummite rims demonstrates a very close location of secondary mineralization near the primary ores, which indicates poor mobility of oxidized uranium. Secondary uranium mineralization occurs directly at the place of primary one, inside the primary ore contour. Uranium transport is probably in the form of

Table 2.2 Stages of redox mineral zoning formation on U-deposits in hypergenesis zone. (According to L.N. Belova's data is compiled by O.A. Doynikova) //Presenting all types of U-concentratings and common features of U mineral-formation in hypergenesis zone: change of mineral associations with changes in environment (Eh, pH) in direction of oxygen waters infiltration (from top to down)

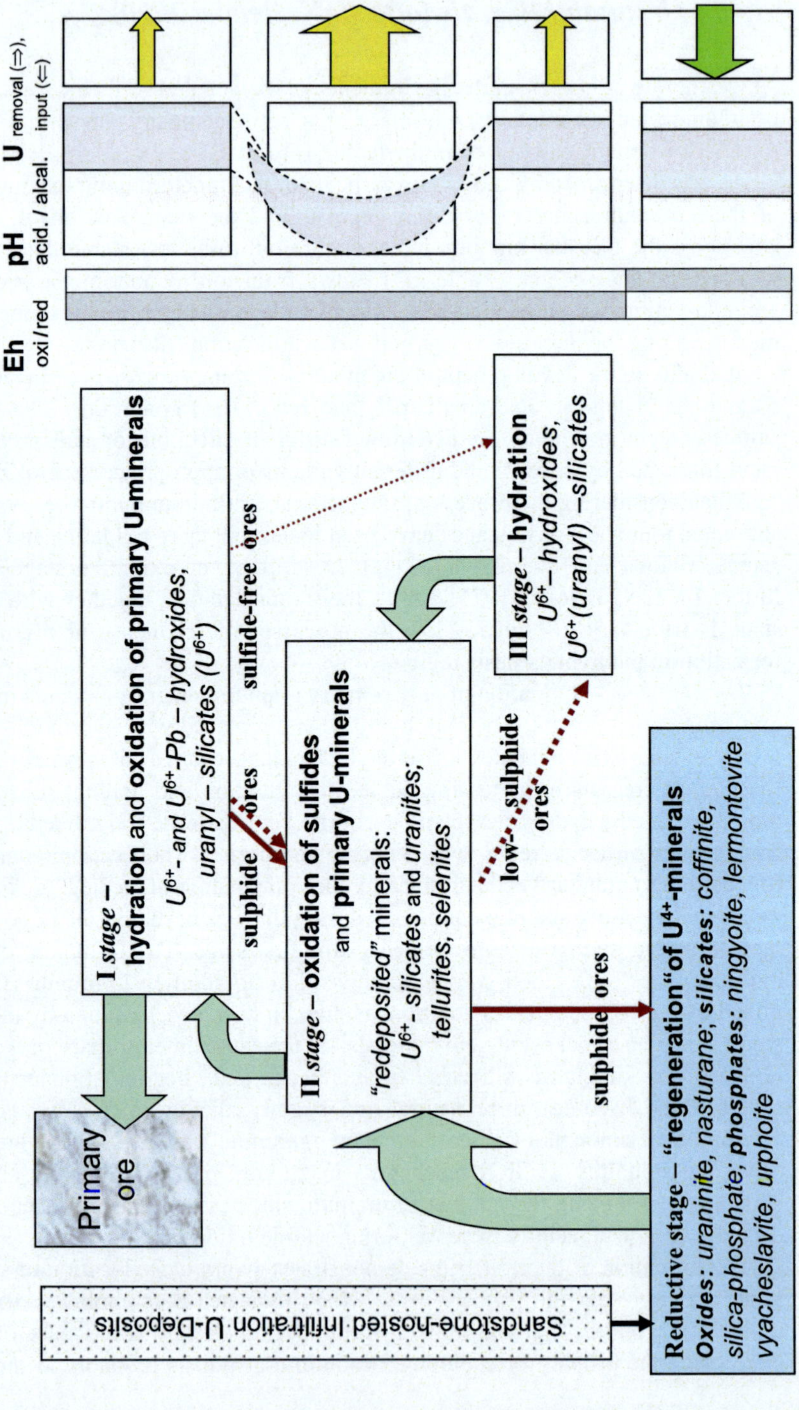

organic complexes due to some amount of fulvic acids, which are invariably present in the soil (Perelman and Kasimov 1999) and participate in the composition of seepage surface waters. In the absence of carbonates and increased uranium concentration (more than 10^{-4} M), it is possible to find in solution uranyl polymers of type $[UO_2(OH)_2{}^{2+}]$ (Hydrogenic... 1980). It can be assumed that at hydration of uranium ores at the first stage of oxidation (as well as in the case of oxidation of unsulphide uranium ores) such uranyl complexes are formed. In this case, satisfactorily explains the variety of mineral and polytypical species of uranyl-hydroxides, which main building unit is identical or similar in structure to composition of uranyl anionic layers $[UO_2(OH)_2]$ (Sidorenko 1978).

Widespread Pb-containing uranium hydroxides may contain lead of radiogenic origin, which gets into mineral-forming solutions at destruction of primary uranium minerals. Then the qualitative composition of uranium hydroxides depends on the time of uranium ores formation. For example, the Proterozoic uranium ore deposit (Shinkolobwe) has a very wide range of complex-cationic uranium hydroxides (masuyite, fourmarierite, curite, vandendriesscheite, wölsendorfite). Sulphide-free Neogene uranium ores (North Caucasus) contain only cation-free and low cationic uranium hydroxides (becquerelite and skupite).

The second stage of the oxidation process begins with the sulfides oxidation, proceeds with acidity degree increasing, at the simultaneous oxidation of sulfides and uranium minerals. Sulfide oxidation provides for appearance of sulfuric acid solutions and, as a consequence, formation of readily-soluble, mobile uranyl complexes ($UO_2{}^{2+}$). Development of this stage depends on the degree of sulfide content of ore bodies (or degree of host rock saturation with reducing agents). With sharp predominance of sulfides in primary ores and, as consequence, in acidic environment and predominantly sulfate solutions, uranium actively removed outside the oxidation zone (in form of uranyl complexes). However, manifestations scale of acidic solutions action may be limited by the presence of neutralizing minerals or carbonate composition of veins and host rocks. Of these two factors, mutually balancing each other, L.N. Belova chose the most general one (presence of sulfides) to describe general regularities; second factor—less common. So we are forced to neglect second factor (presence of neutralizing minerals). But one should always keep in mind the possibility of its influence when carbonates presence. In case of wide manifestation of neutralizers (carbonates), second stage minerals will be shown in spotted areas form, and minerals of the first stage will be corroded only partially.

In ideal case of neutralizers absence (or their complete dissolution before the beginning of sulfides dissolution) in the second stage minerals of the first stage completely (or partially) disappear, because unstable in the acidic environment. In case of wide distribution of sulfides the acidic environment becomes more aggressive; and then first stage minerals can be completely replaced in the oxidation zone. Conditions of the second stage are favorable for uranyl minerals (uranium micas) formation: phosphates and arsenates of autunite type $[Ca(UO_2)_2(PO_4)_2 \cdot 10\text{-}12H_2O]$. The more different sulfides (and arsenides) are present in the ores and the more diverse cation-anion composition of seeping solutions (due to the host rocks composition), the more diverse micas composition.

Emerging redeposited uranyl minerals (micas and silicates) are formed at some distance from the primary ores, indicating the increasing mobility of uranyl complexes at increased environment acidity. With sharp predominance of sulfides (acidic environment), so at predominantly sulfate composition of solutions, uranium is actively removed (in form of uranyl complexes) outside the oxidation zone.

For deposits where sulfides are in subordinate relation to primary uranium minerals, mineralization of the second stage is weak: along with the appearance of uranyl micas, the first stage minerals will be preserved (mica and mica-silicate types of oxidation zone are formed, see Table 2.1).

In case of insignificant amount of sulfides in the primary ores, the second stage falls out completely, the first and the third stages merge. As a result, only a progressive change in the "gummite rims" is observed—their enrichment with silica and alumina.

The third stage—the change of previously formed minerals after complete oxidation of sulfides. It is manifested in the deposition of uranium hydrates (schoepite, paraschoepite), which indicates the alkaline nature of final stage of oxidation zone formation. In terms of scale and intensity, it is subordinate to the second stage. The transition from second to third oxidation stage is clearly fixed by the development of phosphuranilite group minerals [$Ca(UO2)_3$ $(PO4)_2$ $(OH)_2$· $10H_2O$].

Increase in the relative number of uranyl groups in comparison with the autunite and appearance of OH-groups in the composition of these minerals, also indicates the increasing alkalinity of the mineral formation environment. The acidic environment, where uranyl micas were stable, changes to a more alkaline environment, where phosphuranylite and its arsenic analogue are stable.

The reducing stage of hypergenic process, which forms the reducing mineral zonality, is typical for deposits with increased sulfide content of primary ores. This stage is manifested as in crusts of weathering, so in various types of infiltration deposits. In the last it is occurs when presence of reducing agents in ore-bearing strata, represented by sulfides or organic matter. Stage differs from the previous oxidating ones on inputting uranium (instead of removing) and deposition of polymineral black mineralization (Doynikova et al. 2002).

This last stage of hypergenic transformation of primary uranium ores is manifested in formation of cementation zone (secondary enrichment), which is naturally associated with the overlying zones of oxidation processes. Essentially, the secondary enrichment zone represents the lower horizons of hypergenic processes. In the process of infiltration ores formation it corresponds to a kind of cementation zone with reduction conditions, following the oxidized zone, and represented by the richest ore mineralization. In both cases, at a certain reductive interval, so-called "regenerated" uranium blacks are formed, creating sharply increased ore concentrations of uranium.

The full manifestation of all stages of redox process, including the reduction process, is due to the obligatory presence of reducing factors. In addition to sulfides, reductive factor in the development of the hypergenic process can also be the presence of organic matter in the ore-bearing strata. This is especially evident in the formation of black mineralization in various infiltration deposits. As noted above, the ores from these deposits are essentially demonstrating a broadly represented reduction stage of the hypergenic process.

The regularities of uranium mineralization in the hypergenesis zone established by L.N. Belova (1975, 2000) allow to judge about the primary ores composition at the depth and make prediction estimates. For example, in sulphide-free deposits, where uranium is hardly ever transported outside the ore bodies, it is possible to quantify the scale of primary mineralization by secondary oxidation zones. The study of oxidation zones of sulfide-uranium deposits, where dilution of ores is typical, allows us to judge about the possible formation of secondary enrichment zone, which is of industrial interest, and to judge about the scale of its development.

2.1.3 General Regularities of Uranium Mineral-Formation in the Hypergenesis Region

The scheme (Table 2.2) shows the general regularities of uranium mineral formation in the hypergenesis region, showing a consistent change in the mineral-formation medium (Eh and pH) in the direction of oxygen water infiltration (filtration) (from top to bottom). Thus, the scheme shows all types of uranium concentration in the hypergenesis region and united by common regularities of mineral formation.

When comparing the peculiarities of development of primary ore weathering processes (crust formation) and such of infiltration ore formation processes, the following common features are revealed.

- Acidification of environment recorded in appear of U mica (second oxidation stage of primary ore) corresponds to oxidizing conditions of ore-bearing stratum (pre-ore zone) in stratal-infiltration deposits.
- Subsequent increases in alkalinity (pH) of mineral formation environment immediately before the secondary enrichment zone, was established in oxidation zones of primary uranium ore on the characteristic formation of phosphuranilite group minerals (third stage). In infiltration deposits increases pH is traced by the results of studies of iron minerals along the ore-bearing layer. Both montmorillonitization of pre-ore intervals and goetitization of clays on the boundary of ore body with oxidation zone testify to the alkaline environment. In localized here yellow clays, unlike the grey clays from the unoxidized and fully oxidized parts of stratum, montmorillonite was fixed according to the X-ray phase analysis (Hydrogenic... 1980).

For pre-ore intervals of infiltration deposits, the characteristic Fe removal from the front part of the plast oxidation (subzone of "partial oxidation") was noted earlier (Kashirtseva 1970). Corresponding whitewash of rocks is treated as redistribution (removal) of iron under conditions of local gley conditions (Hydrogenic... 1980). This phenomenon was detected by observations of the "whitewash" zones in ore-bearing formations immediately before the ore interval (in the direction of filtration).

Therefore, the regularities of changes in the environment of mineral formation, identified by L.N. Belova in zone of uranium ore oxidation by the change of uranium mineralization, are traced also in the stratal oxidation zones of infiltration deposits (on analysis of various iron forms distribution), which allows to combining them in a common table (Table 2.2). First of all, we mean the character of the pH change in the direction of seepage, which in this case characterizes the course of the hypergenic processes more visually than the change of Eh-potential values.

Generalized consideration of uranium mineral formation process in hypergenesis region implies study of both oxidation zones (oxidizing environment) and closely related secondary enrichment zones, where geochemical environment changes on reducing one.

The redox zoning of uranium hydrogenic deposits has been considered in detail earlier in a number of Russian monographs (Kashirtseva 1970; Hydrogenic . . . 1980; Kislyakov and Shchetochkin 2000) where, based on an extensive field material, the mineralogical-geochemical zonality of plast-infiltration uranium deposits is in detail considered and scheme is given. The distribution of different valent forms of iron in this plast infiltration deposits was analyzed. As a result, stratal oxidation zone are divided into subzones of "full", "incomplete" and "partial" oxidation. The selection of "ore body" subzone and of outrunning "halo of uranium scattering" in the reduction zone inherently indicates the presence of secondary enrichment zone (ore body). It is lawful to assume that such zonality corresponds to the geochemical laws of uranium mineral formation, which were observed by L.N. Belova in uranium deposits oxidation zones.

About inseparable unity of redox zonality in the weathering crust of uranium deposits was repeatedly stated in the works of outstanding Russian soil scientist and geochemist A.I. Perelman (1916–1998). Famous scientist on uranium geology N.P. Laverov (1930–2016) repeatedly underlined that geochemically contrasting oxidizing and reducing conditions must be spatially connected with each other. Therefore, when considering from a single position the conditions and regularities of uranium ore formation in all manifestation in hypergenesis region, it is advisable to talk about *united redox* zonality which includes oxidation zones of U deposits as component part (Belova and Doinikova 2003b).

Regardless of scale of reduction stage manifestation—cementation zone or local areas in sedimentary rocks—it is always represented by the so-called blacks' mineralization (black type zonality in Table 2.1). Study of this dispersed uranium mineralization in different types of infiltration deposits, including secondary enrichment zones, allowed us to speak about the similarity of the conditions and regularities of the blacks' ore formation.

2.2 Examples of Uranium Deposits with Commercial Black U Ores

Uranium black is widely manifested in sedimentary mantle deposits and in secondary enrichment zones of hydrothermal (endogenous) uranium deposits. In hydrogenic deposits uranium black is of the greatest industrial and scientific interest. As noted above, such deposits are promising for commercial development by the method of underground in situ recovering (ISR). The study of such ores mineralogy helps to find out the conditions of their formation.

Let's briefly consider examples of various genetic types of uranium deposits, including sandstone type, where black mineralization is developed in scales of industrial interest.

First, let's clarify the terminology used below. The term "exogenous-epigenetic", traditionally used in Russian uranium geology to define deposits in sedimentary mantle, is an archaic term at the present level of geological exploration degree of such deposits. Introduced in the middle of the last century, the term rightly pointed out that mineralization occurred after sedimentation, rejecting the "sin"-genesis (simultaneity) of ore formation, which was discussed at that time along with the "epi"-genesis of ores. The first part of the term discussed refers to the subsurface nature of the ore-forming processes; while the second part—"epigenetic"—is not characterize own ore-formation process (compared only the periods of sedimentation and ore-formation).

The modern term "hydrogenic", which is used to describe sedimentary mantle deposits, seems to be more precise, giving the genetic characteristic of the proper ore-formation itself. The review monograph "Hydrogenic ore formation" (Kislyakov and Shchetochkin 2000) fully represents the modern systematic of hydrogenic deposits (at first including uranium deposits), which is accepted in the present work. Let's remind that the term "hydrothermal" now concerns deposits which are connected with high-temperature deep processes.

At hydrothermal deposits, uranium blacks are formed in the process of weathering of sulfide-containing uranium ores in the lowest zone of hypergenic processes, known as the secondary enrichment zone. Under the influence of oxygenated surface waters, pseudomorphic substitution of primary ores takes place with the formation of regular redox zonality as described above. The scale of such transformations, as well as the development of the secondary enrichment zone itself, depends on a number of geological, geochemical and hydrological factors. The same process of U-ore formation is in the infiltration sandstone deposits. When uranium-containing oxygen surface waters enter the sedimentary strata, uranium black ores of sandstone-type deposits are formed.

2.2.1 *Examples of Complete Substitution of Primary U Ores in Hypergenic Environment (Taboshar and Tuya-Muyun Deposits, Central Asia)*

The Central Asian deposits, Taboshar and Tuya-Muyun, are examples of complete substitution of primary ores in hypergenic conditions.

At Taboshar deposit, the primary ores are entirely replaced by uranium blacks. The deposit is located in Karamazar ore district, Central Asia, in the southern part of Kuramin Ridge (western periphery of the Tien Shan). According to the data (Rybalov et al., IGEM funds, 1956), this deposits can be considered as linear stockwork of complex fracture-vein zones, concentrated within area of ~10 km sq. within the intrusive rocks (graniteporphyries, penetrating granodiorites).

Taboshar is a deposit, where for the first time in the Soviet geology in the 40s a black type of commercial uranium ores was discovered. Schematically, this deposit is an ore body, deeply processed by oxidation processes, fully localized in the tectonic disturbances zone of the host rocks. During first years of this deposit development it was noted: *"The modern appearance of ores is determined by an exceptionally intensive development of oxidation processes, which was facilitated by a strong tectonic remaking and sulfides presence"* (Gritsaenko et al., IGEM founds, 1956). Essentially, all ore is composed of secondary uranium minerals (*"U-mica, residual and re-deposited blacks"*), and primary uranium formations *"are of mineralogical interest only*". The oxidation zone presented by the U-micas extends to 20–30 m depth (groundwater level) and is combination of mineralization of last oxidation stages. This indicates the complete disappearance of the first stage oxidation minerals as a result of their replacement under the influence of acid solutions caused by the presence of sulfides. Further to ~300 m depth from the surface, in zone of saturation with fracture-ground waters, where there is no oxygen, black ores of the secondary enrichment (cementation) zone are localized (Fig. 2.3).

Uranium (U-V) deposit Tuya-Muyun located in the southern frame of Fergana Valley in Central Asia should be attributed to the same type of deposits as Taboshar, according to character of deposit localization in modern weathering crust, when the whole ore body completely processed by oxidation processes. According to A.E. Fersman's ideas, the deposit was formed as a result of deep circulation of underground waters, that borrowing U and V from coal-siliceous shales (Kislyakov and Shchetochkin 2000). This deposit's ores are composed exclusively of secondary uranium minerals, i.e. U-micas colloform formations. Unlike the Taboshar ores, there are no black ores, i.e. no secondary enrichment zone.

The study of Tuya-Muyun ores matter composition has always been complicated by the collomorphic fine-grained nature of the ore aggregates (dripstone). For a long time, since the discovery deposit in 1925, the question of the primary uranium ore composition remained open. Our research of ore samples from the lower horizon of this deposit with the help of ATEM methods allowed us to determine the primary ore mineral: it is dispersed, so-called "residual", nasturane (Belova et al. 1985a).

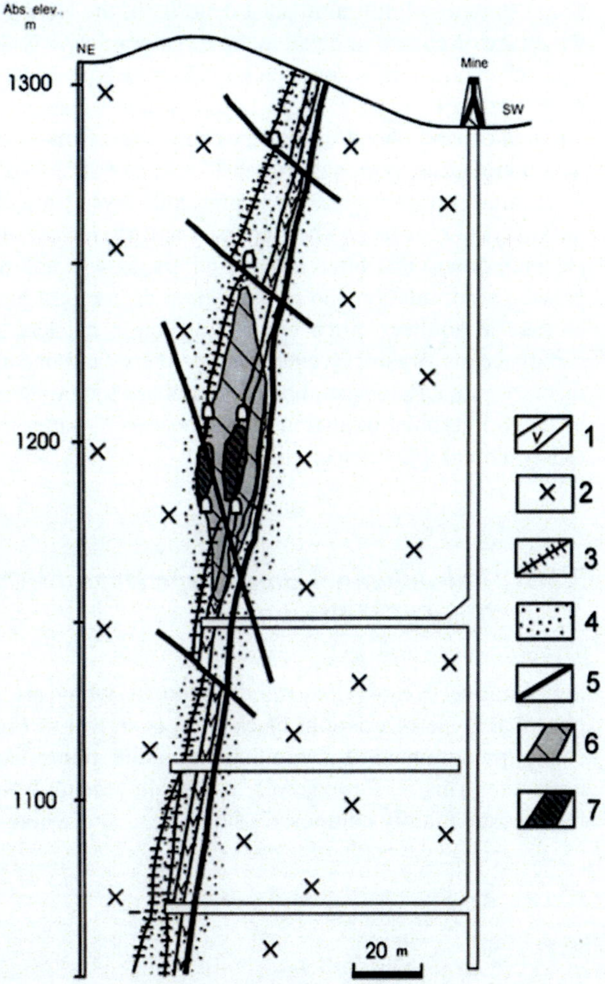

Fig. 2.3 Schematic geological section of the Taboshar uranium deposit, Central Asia, with commercial black ore-mineralization/ according to (Rybalov B.L. et al., IGEM funds, 1956): (1) diabase porphyrites, (2) granite-porphyries, (3) barite and quartz veins, (4) zones of beresitization, (5) tectonic disturbances, (6) ore body; (7) rich ores

According to the laws of uranium deposit oxidation zones formation (Belova 2000), based on the oxidized ores mineralization character, we can judge about the type of primary uranium ores. In this case, the absence of secondary enrichment zone at the deposit clearly indicates the absence of sulfides in the primary ores composition. An interesting mineralogical conclusion follows from this: the primary ores of Tuya-Muyun were properly uranium (nasturane like) and non-sulphide ones.

2.2.2 *Stratum-Infiltration Deposits (Chu-Sarysu Type)*

A typical example of large industrial facilities that are promising for commercial development by the ISL (or ISR—in situ recovery) method is the rich (often very large) stratum—infiltration plast deposits of the Near Tien Shan ore mega-province, which are discussed in detail in the monograph (Kislyakov and Shchetochkin 2000). The schematic section of such Chu-Sarysu type hydrogenic deposit (Fig. 2.4) is a combination of plast-like ore bodies limited inside subartesian highly permeable aquifers of sand-clay sedimentary complexes. Roll—type morphology of ore bodies are controlled by regional or "local" zones of stratal oxidation. As a rule, the source of uranium inflow is basement granitoid crystal massifs, which usually come out onto daylight surface. They are washed by surface waters which then penetrates (seeping down) and filtered into aquifers. Low-uranium-bearing oxygenated waters provide accumulation and re-deposition of uranium black' ores on mobile reducing barriers in aquifers. Such ore-sedimentation barriers are due to the abundance of carbonaceous organic (vegetable) residues (and/or sulphides) in productive sedimentary strata. Recent studies show that ore accumulation on such a reducing barrier is due to microbial oxidation of carbonaceous matter in the host gray-colored rocks (Kislyakov and Shchetochkin 2000).

2.2.3 *Paleochannel Basal Type Deposits (New Phosphate Type of U Black Ores)*

Schematic section of *paleochannel* deposits, which are characterized by little-known phosphate type of uranium black ores, is shown in Fig. 2.5. Their main or significantly predominant ore component is the phosphate of tetravalent uranium—ningyoite. This is, for example, paleovalley deposit Ningyo-Toge in Japan, where the ores are mainly composed of ningyoite and where this U^{+4}-phosphate was first

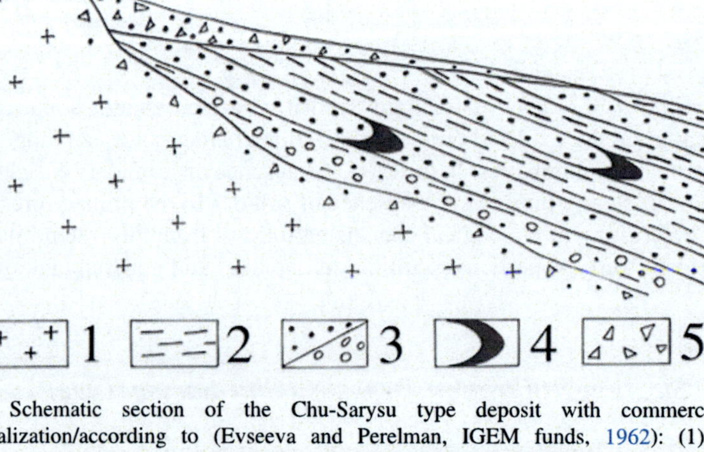

Fig. 2.4 Schematic section of the Chu-Sarysu type deposit with commercial black ore-mineralization/according to (Evseeva and Perelman, IGEM funds, 1962): (1) granites, (2) clay rocks, (3) sandy rocks with pebbles, (4) ore deposits, (5) gravelites

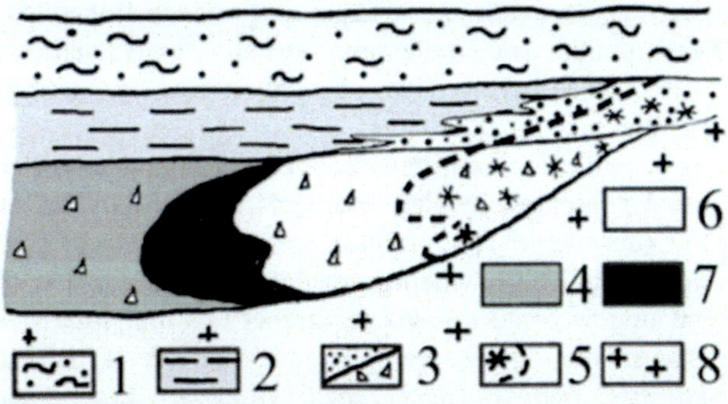

Fig. 2.5 Schematic longitudinal section of basal (paleovalley) type uranium deposit with commercial black ore-mineralization/according to Kislyakov and Shchetochkin (2000): (1) loose deposits, (2) clayey rocks, (3) sandstones and gravelites, (4) primary gray-colored rocks, (5) variegated-oxidized rocks, (6) bleached rocks, (7) uranium mineralization, (8) granites

discovered. Bulgaria deposits, where ningyoite is the only (Momino, Navysen) or the main (Maritsa, Haskovo) ore component belongs for same type. Rich ore samples are loose (sooty) material of black to dark brown color, i.e. mineralization quite corresponds to the name "blacks". In recent years ningyoite ores was finding at Vitim paleochannel type (paleovalleys) deposits in Russia (see Chap. 5).

When studying uranium ores from infiltration deposits in Bulgaria, two types of deposits and two forms of phosphate-ore mineralization have been identified (ningyoite and ningyoite—pyrite). In all cases this ore mineralization is finely dispersed, and is localized among non-oxidized rocks directly near the contact with oxidized rocks. Ore-hosting rocks of the first type deposits (Momino, Navysen) are *paleovalley/paleochannel* sedimentary strata of Pliocene age, which correlates them with the similar uranium deposits of Japan and Canada (Boyle et al. 1981). Ore minerals in such deposits are frequently pure ningyoite without impurities. The second type of deposits (Haskovo, Maritsa) is confined to the Upper Eocene tuff-sedimentary strata. Ore bodies with predominantly ningyoitic mineralization (the amount of coffinite and uraninite is sharply subordinate) are located in the gray-colored strata at the contact with green-colored ones. Green-colored rocks usually underlay intercalations of gray-colored rocks, but sometimes in lithologically same layer there are gradual transitions of colors or their cutting contacts. According to mineralogists of IGEM (fund material), green-colored rocks are the result of the reduction of previously oxidized limonitized interlayers. The ore mineral in the second type of deposits is represented exclusively by Fe containing ningyoite. Younger, typically roll-shaped formations were also found in this type of deposit at the contact of oxidized and unoxidized tuffs.

According to classification (Kislyakov and Shchetochkin 2000), uranium deposits in paleovalleys belong to the phreatic (subsurface) infiltration subclass. In contrast to the stratum-infiltration deposits, they are formed as a result of free non-pressure groundwater filtration, and are characterized by shallow depths (up to n-100 m). Here ore reserves morphology are similar to that in plast-infiltration ore deposits. Such paleovalley/paleochannel deposits are confined to permeable sandy alluvial deposits of ancient valleys, the channels of which embedded into crystalline basement. The uranium mineralization, usually associated with the upper paleovalleys, was formed by the flows of uranium-bearing freatic (ground) waters directed towards the local erosion bases.

Probably, geochemical conditions of alluvial, organically rich (and, consequently, microbial active) paleovalley sediments, are the most favorable for formation of phosphate (ningyoitic) type uranium ores. It is because of given conditions of free water exchange and shallow depths available for atmospheric oxygen.

Our data and review of all known occurrences of ningyoite (Doinikova 2007) have shown that mononingyoite composition of U-ore is the characteristic for paleovalley deposits: Ningyo-Toge, Japan; Blizzard, Tyee, Canada; Momino, Navysen, Bulgaria. Uranous phosphate ores of Vitim plateau in Russia are also included in this series (Doynikova and Tarasov 2014). Besides ningyoite, coffinite and nasturane (amorphous uraninite) are in subordinate amounts in ores of the infiltration type deposits (of stratal—and ground—filtration). Such composition is in the economic secondary ores of the Grachevskoye ore field (Kosachinoye); in ores of stratiform deposits of the Czech-Bohemian massif and in ore-bodies in tuff-sedimentary strata of Bulgaria (Navysen, Maritsa). In case when uraninite is predominant over ningyoite in ore composition (Sugraly, Kyzyl Kum) we have not seen carbonaceous organic residues in sandstones. By the 2000s, it became obvious that the deposits with ningyoite ores are not uncommon, as previously thought, but represent a new type of phosphate black uranium mineralization, characteristic of the paleo-channel deposits.

The composition of phosphate-black U-ores demonstrates increased PO_4-ion activity during their formation. The biogenic nature of the main mineral-forming element P detail discussed future in Chapter 5. Supposed prevalence predominance of biogenic contribution to such uranium blacks' genesis is confirmed by both geological observations (the presence of biologically active lacustrine-bog sediments) and mineralogical data (phosphate composition of uranium ores).

2.2.4 Ground-Infiltration Weathering Deposit (Secondary Ores of Kosachinoye)

An example of commercial scale uranium black's accumulation in the weathering crust is secondary ores from the Kosachinoe deposit of Grachevskoye ore field, Northern Kazakhstan. It will be considered in more detail than others earlier discussed because of the sharply different characteristic of hypergenic transformation of primary ores. All used data on geology and mineralogy of primary ores is taken from unpublished materials (Geologic reports, IGEM funds, 1988, 1977); secondary ore mineralization has been studied by us later using ATEM methods.

This deposit is accumulations of slab-shaped, flatly elongated, or columnar metasomatic ore bodies mainly confined to linear fault zones at the contact of

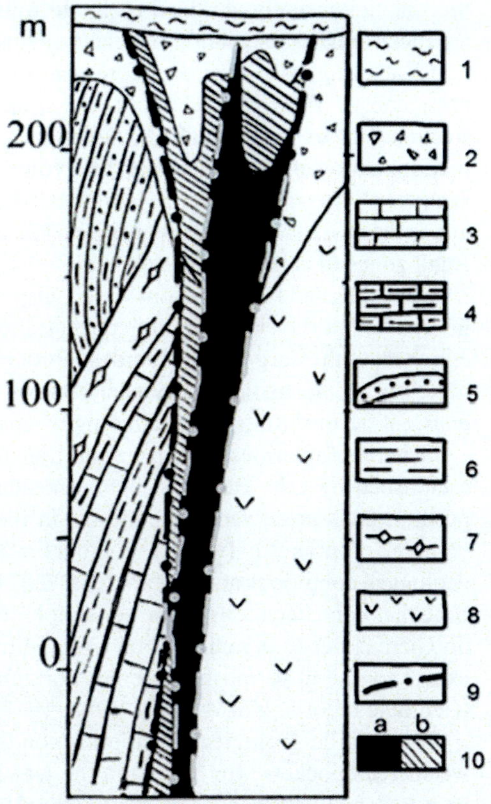

Fig. 2.6 Schematic geological section of uranium deposit Kosachinoe, N. Kazakhstan with commercial black secondary ores/according to (Gorshkov et al. 1988): (1) loose deposits, (2) weathering crust formation, (3) limestones and dolomites, (4) carbonaceous limestones, (5) quartz sandstones, (6) clay shales, (7) zones of tectonic melange, (8) basalt porphyrites, (9) faults, (10) ore body rich (**a**) and poor (**b**) ores

sedimentary strata with intrusive (subvolcanic) bodies, or to zones of intense crushing and increased fracturing at the intersection of large faults.

Ubiquitously developed in hypergenesis region, intensively worked out, mature weathering crust up to 150 m thick is composed of clay alluvial formations (loose yellowish-gray mass). Hypergenic formations can be traced by cracks to 250 м. Above ore zones, where tectonic disturbance of ore bodies is high, the formation of weathering crust "pockets" is typical, with increase in capacity to 1.5 times. Industrial mineralization is localized in the weathering crust, where there is a noticeable increase in the capacity of ore bodies and their enrichment (up to n%) compared to the primary proper uranium hydrothermal ores in the cracks at the depth.

The typical Kosachinoe schematic section of the slab-like ore body on the contact with the basalt massif is shown in Fig. 2.6. The ore body of flatly elongated shape (with blowing up to 10 m), keeping the steady continuation along the contact, is traced in the weathering crust (with some form washout) from deep to the sole of the uppermost multicolored horizon. The primary U mineralization is represented by nasturane and X-ray-amorphous coffinite and U-containing titanate (presumably brannerite). Uranium formations of thin (n·mm) veins or cementing material brecciate intensively albitized rocks, forming massive blocks of ore bodies, which are metasomatic association of uranium-bearing albitites.

Weathering crust ores are represented by secondary black formations. In these ores, consist from clay eluvium, the bulk of uranium is concentrated in pelitic fraction (<0.01 mm). ATEM studies have shown that these ores are composed predominantly of Fe-containing ningyoite. Lumpy coffinite is less common. All samples always contain poorly crystallized uraninite; it is often electron amorphous and is diagnosed only by composition. The richest ore areas are confined to peculiar small zones of secondary enrichment.

According field observations, finely dispersed uranium mineralization is localized in organic interlayers. Uranium concentrations are maximal in coal-siliceous shales enriched with Corg and sulphides. Enrichment of ores in the weathering crust indicates the formation of typical zone of secondary enrichment, which usually does not occur during the weathering of unsulphide uranium ores.

General regularities of hypergenic transformations of non-sulfide uranium ores, established by L.N. Belova (2000), are manifested on Kosachinoye in ore bodies position (it is preserved and continue in the weathering crust), in the insignificant removal of uranium (weakly contours washout). However, the combination of geological conditions at the deposit has led to the formation of secondary enrichment zone that non characteristic for non-sulphide (low-sulphide) uranium ore. Its formation here is due to organic carbon Corg which is widespread in the weathering crust except for the uppermost multicolor horizon, where the oxidative situation prevails (Geologic reports, IGEM funds, 1988, 1977).

The specific geological situation is conditioned by the development of powerful weathered "pockets" (to 250 м) in the weathering crust over productive areas with high tectonic disturbances of ore bodies. Here, deep areas of stagnant oxygen-free environment, favorable for recovery conditions, are created in the hypergenesis region.

The humid character of conditions in which existing weathering crust was formed is indicated by the increased content of beryllium in ore intervals (up to 5 Clarks), which mobility in hypergene process is higher in humid landscapes (Perelman and Kasimov 1999). The beryllium trend of accumulation in wetland soils characterizes to some extent the landscape conditions of Be accumulation in this deposit. Conditions of migration and accumulation of beryllium are in many respects comparable with the conditions of Kosachinoe weathering crust. That suggests the presence here of alkaline phosphate barrier precipitating for beryllium in the hypergenesis region. Predominantly phosphate (ningyoite) composition of secondary ores, as well as rare manifestations of phosphate U-mica, demonstrates these conditions of increased concentration of phosphate ions. In this case, the entire area of ningyoite mineralization is the geochemical barrier. Weakly acidic and neutral pH values, typical of ningyoite formation, correspond to the conditions which necessary for beryllium deposition.

The Kosachinoye deposit is also unusual in the form of ore body shape in the secondary enrichment zone. In this case, it is difficult to call it a cementation zone, because among the loose yellowish-gray weathering crust deposits ore intervals differ only by darker shade of gray. This zone occupies predominant part of the weathering crust under multicolored horizon of quaternary sediments. The loose industrial ores localized here differ from the loose weathering crust formations (without uranium) only by the presence of uranium minerals and an increase in the content of C_{org} and sulfides.

According to geologists studies (IGEM funds, 1988), the C_{org} content are decreased in primary ores and in near-ore metasomatites in comparison with the amount of it in the host rocks. This was explained by its «combustion», i.e. decomposition during the oxidation of organic matter in the process of ore formation. Body morphology of primary ore is traced into the upper stratum of hypergenesis region as its continuation, where ore body is enriched with organic carbon (up to 15% C_{org}) and sulphides as compared to the primary unchanged ores. Geologists' conclusion about accumulation of C_{org} "in situ" due to the inflow of surface water from the soil layer is logically completing, if assume the leading role of biogenic factor in the process of secondary enrichment.

The increased sulphide content of loose ores is likely to indicate the activity of sulfate reducing (desulphurizing) bacteria. These bacteria develop in biosphere systems which inaccessible to free oxygen and contain organic matter scattered in sedimentary rocks. Anaerobic bacteria take away oxygen from sulfate ions to oxidize organic matter, which creates conditions for sulfur reducing to H_2S and the formation of biogenic sulfides (Perelman and Kasimov 1999).

As sulfides formation due to decomposition of the C_{org} (previously assumed), so the presence of C_{org} and sulfides along the entire profile of weathering crust can be convincingly substantiated by bacterial activity. Apparently, the formation of secondary ores in this deposit was largely facilitated by the microorganism community, inherent in all weathering formations, which is particularly diverse in the case weathering of sedimentary rocks containing plant material (here in form of frequent carbonaceous interlayers).

The conclusion about the significant role of biogenic factor in Kosachinoe deposit commercial ores formation can be supported by the following research results. Preferable confinedness of Corg to ore intervals can be explained by the presence of sulphate sulfur in primary ores (in form of post-ore gypsum and barite occurrence), which is used in sulfate-reducing processes of bacteria activity. In addition, it is not excluded the presence of other anaerobic microorganisms known in bacterial hydrometallurgy, which life activity associated with uranium. That leads to conversion of uranium mineral forms into organic compounds. In this case, ores enrichment with organic carbon in the weathering crust is logically explained by increase in bacterial biomass as a result of bacterial activity.

The extremely friable nature of commercial ores indirectly indicates the activity of bacteria. Another indirect indication of microorganisms' widespread activity in hypergenesis process can be mainly clay composition of crust formations. The absence of carbonates widely represented in bedrock can be result of autotrophic bacteria activities that extracting and processing carbon from inorganic mineral forms; in this case, from sedimentary carbonates. It is likely that autotrophic microbial communities (including sulfate-reducing bacteria) played a role of reducing factor in oxygen-free subzone of Kosachinoye deposit.

On the example of this deposit, it is observed significant *influence of climate* on the features of U-mineral zonality formation in the weathering process of primary U-ores (see Chap. 5). Thus, the regularities of uranium deposits hypergene transformation, repeatedly observed in the conditions of arid climatic zone of Central Asia (L.N. Belova's systematic has been developed at this type deposits), are supplemented with new features in the case of significant contribution of humid component to climatic conditions. Thus, in continental climate (Kosachinoye), the abundance of organic matter in weathering crust in combination with favorable geological conditions can provide the formation of a secondary enrichment zone, which is not typical for proper-uranium ores type.

2.2.5 Participation of Microbiological Factor in Black's Uranium Ores Formation

Summarizing results of this brief geological section, it should be noted that in the study of black uranium mineralization increased attention is always attracted by biogenic factor of uranium ore formation. Geologists' interest in biogenic features of uranium-bearing stratum sediments, which arose in the 70s, is reflected in the constant attention to this topic, and it is discussed in generalizing monographs (Hydrogenic… 1980; Rare-metal… 1995; Kislyakov and Shchetochkin 2000). *"The biogenic factor manifests itself as a generator of reducing conditions"*—so academician N.P. Laverov briefly formulated the formation theory essence of so-called exogenous-epigenetic (hydrogenic) uranium deposits.

Following V.I. Vernadsky, who noted the significant role of bacteria in formation of manganese ores, "the outstanding role of organic (bacterial) world during the diagenetic transformation of sediments" was noted by N.M. Strakhov (1957). The microbiological nature of a number of iron and ferromanganese ores is now well known. Soil formation caused by bacterial activity is considered as a model of weathering and crust-formation processes (Perelman and Kasimov 1999). In particular, bacterial activity is associated with formation weathering crust of sulfide deposits. The important role of microorganisms in formation of reductive geochemical barriers that localize uranium ore accumulation was first declared by A.K. Lisitsin (Lisitsin and Kuznetsova 1967). Modern hypothesis of biogenic formation uranium deposits (Vinichenko 2004) supports Vernadsky's ideas about role of biosphere in the Earth's crust formation, relying also on scientific investigations of Russian scientists—A.K. Lisitsin, G.A. Zavarzin, A.I. Perelman and others. When analyzing materials on study modern uranium accumulation in peat lands and in coal shale, Siberian field geologist V.P. Vinichenko came to conclusion: "U-ore formation depends on the activity of living microorganisms, and is not subject to chemical laws". The recognition of the living matter leading role in uranium ore accumulation allows him to declare that uranium matter concentration in aqueous (hypergenic) solutions occurs as a result of biogenic processes caused by living bacteria. He convincingly explained conditions for rich ores formation on deposits of unconformity type. "Due to the vital activity of aerobes producing an abundance of organic acids (in oxidizing zone), rich U-ore-bearing solutions is formed, in which the uranium content reaches hundreds of mg/liter". Aerobic metabolism products, in turn, provide anaerobic activity. In the opinion of this theory author, aerobic bacteria, form (!) reductive environment, and do not increasing reducing properties of the environment, as it is accepted in modern geochemical views on uranium ore formation (Hydrogenic . . . 1980, etc.).

According to (Vinichenko 2004) « . . . in existing genetic models, only role of anaerobic bacteria producing hydrogen sulfide, hydrogen, methane is taken into account. The "work" of aerobic microflora in the barrier formation remained almost unnoticed». The increase on bacterial activity, what has been repeatedly noted earlier, is not consequence of geochemical barrier presence, but a cause of its formation. It is such reductive environment that is *biochemical,* rather than *geo*chemical, barrier that precipitates uranium. "The structure of the ore-forming barrier organically includes not only reducing but also oxidizing zone; . . . the ore-forming environment is consisting of the redox *bio*chemical (biogenic) barrier". Such look at the U-precipitating ore-forming barrier is broader than it is now accepted; it complements the existing theory of exogenous ore formation.

The leading role of microbiological processes in creating a reducing environment in sedimentary strata is confirmed by the results of modern research on environmental mineralogy. Various geological publications recent years also indicate mineral-forming role of microorganisms in the formation of coffinite and uraninite in sandstone type uranium deposits in China and Australia (Min et al. 2005; Wülser et al. 2011; Cai et al. 2007; Zhao et al. 2018; etc). Biogenic genesis of uranium black mineralization will be discussed later in connection with the consideration of uranous phosphate genesis in Chapter 5.

Obviously, the biogenic factor generating black uranium mineralization deserves more systematic study. Further fundamental study of microorganisms' role in the formation of various commercial deposits, especially in industrial important sandstone hosted U-deposits, will undoubtedly contribute to improving the efficiency of geological exploration work.

2.3 Mineral Composition of Uranium Blacks

The mineral composition of uranium blacks must be known both to determine their genesis and to select effective schemes for their processing (leaching) in the extraction of uranium ores from sandstone type deposits.

Until recently, the term "uranium black" in geological literature was usually identified with the only mineral form of uranium—oxide. Such friable sooty black formations are difficult both for optical diagnostics and for diagnostics by traditional mineralogical methods (XRD, IR and X-ray spectroscopy, etc.). All these methods analyze total sample composition (finely dispersed phases mixture), and results characterize the dominant phase in the sample, omitting as sharply subordinate phases so X-ray amorphous. Taking into account the dispersion of the studied minerals and, often, their small size in hypergenesis region, transmission electron microscopy (TEM) was the most effective for study of phase composition of such complex uranium ores. Small number of one of the phases in the polymineral mixture or extreme dispersion substance degree (beyond X-ray's reach) does not obstacle this method.

Since the 70s, together with the famous uranium mineralogist, expert on the U-deposits oxidation zones Dr. L.N. Belova, we studied in detail the fine-dispersed uranium ores from various deposits of sedimentary cover, as well as from secondary enrichment zones of hydrothermal deposits. Our first publications have been devoted either to individual mineral findings or to the crystal chemistry of dispersed minerals found in composition of uranium black by TEM methods. Our conclusion on the polymineral composition of uranium blacks have been published only in short materials of international and national mineralogical forums (from 1984 to 2002). The methods of uranium Deposits in sedimentary strata studying are reviewed in methodical chapter of monograph (Belova et al. 1985b). Our results of uranium mineralogy study by ATEM methods first are presented in review article (Doynikova et al. 2003).

Further here, the results of near 30 years studies of black uranium mineralization are summarized in order to identify the genesis of uranium sandstone-type deposits in the sedimentary cover.

2.3.1 *Brief Historical Background on "Uranium Black" Study*

The term "uranium black" was first introduced in Russian mineralogy by V.I. Vernadsky in 1927 as literal translation of German term "Uranschwärze", which was used by miner workers in the ore mining at famous Jáchymov deposit (Czech Republic) back in the nineteenth century. The term refers to uranium-rich loose sooty formations accompanying *uraninite* and *nasturane*—uranium *pitchblende*. Back in 1938 V.I.Vernadsky suggested that uranium black is product of nasturane decay (possibly radiochemical) below the oxygen surface, above which only UO_3 compounds are forming (Vernadsky 1954).

In 1939, uranium black was selected by V.G. Melkov as a new (independent) uranium mineral form in uranium oxides group (Melkov and Pukhalsky 1957). Proposed by him, separation of two types of uranium blacks—"residual" and "regenerated"—stimulated the study of mineral formation processes in secondary enrichment zones. Attempts to distinguish these two types of uranium black by particle shape in transmission EM-image (suspension preparation) were made in

1958 (G.S. Gritsaenko, IGEM fund). These were first steps in uranium ores study using the TEM method.

Even before the mid-50s, uranium blacks were unambiguously and undoubtedly considered earthy uranium oxides. After coffinite discovery in 1955, wide development of this U^{4+}-silicate in U-ores composition of most uranium deposits was established. Probably, first discovery of coffinite in unoxidized parts of exogenous Colorado deposits, and then its frequent discoveries in composition of hydrothermal (endogenous) ores, led to fact that coffinite was not linked so closely to uranium blacks.

By the middle of the 1970s, the coffinite had already been marked into black's ores composition (Dymkov Yu 1973); summarizing the extensive material on uranium pitchblende mineralogy, herein it is notes the illegitimacy of considering uranium black as a "mineral". He classifies uranium blacks in the same category as uranium pitchblende. "...The mineral composition of blacks is the same: nasturane, coffinite, uraninite...»; uranium blacks is a type of uranium pitchblende, typical for oxidation zones and exogenous deposits. Here coffinite have been considers to be endogenous mineral; it is noted that in some uranium black X-ray studies (including those associated with hypogenic hydronasturanes), it have been identified as coffinite. However, until the end of 1980s, mineral composition of uranium blacks was usually identified exclusively with oxide uranium form, and coffinite presence was considered to be rare feature of such uranium ores. Only in the monograph (Kislyakov and Shchetochkin 2000), where a wide range of issues of hydrogenic ore formation are considered, coffinite is mentioned as a common component of black ores along with nasturane. In modern foreign literature, such ores are called «sooty uraninite» despite of their polymineral nature. Long-term crystallochemical study of black uranium ore on different samples by a complex of APEM methods indicate the multiphase of it U-component and expand the list of uranium minerals in it (Belova et al. 1991; Doynikova 2012a, b; Doynikova et al. 2002), that confirm conclusion of Yu.M. Dymkov, said above.

2.3.2 *Results of Uranium Black Mineral Composition Studies*

The results of our researches show that ore composition of uranium blacks can be represented by at least three different mineral forms: oxide (uraninite of different degrees of crystallinity), silicate (coffinite) and phosphate (ningyoite). That was shown using local methods of transmission AEM (Belova et al. 1985b; Belova et al. 1991; Doynikova et al. 2003). Before using local EM methods, the composition of uranium blacks was usually, as a rule, interpreted by geologists as an amorphous oxide form of tetravalent uranium. In the ores of sedimentary cover uraninite UO_2, coffinite $USiO_4$ and uranous Ca-phosphate ningyoite can be present in different proportions.

According to our observations, in black ores the size of U^{4+}-mineral individuums does not exceed micrometer values; collomorphic depositions and crystalline accumulations rarely reach the values of n·10 microns. All tetravalent uranium minerals composing uranium blacks are characterized by micron/submicron sizes of microcrystals or gel-like formations (metagel).

Minerals of U^{+4} are allocated on rounded grains of quartz (uraninite), on destroyed grains of feldspar and mica (ningyoite), as impregnation of carbonaceous fragments, debris (coffinite, uraninite, and ningyoite). According to the electron diffraction data, the crystalline uranium oxide formation occurred on equally with its amorphous phases; however, unlike coffinite and ningyoite, uraninite was not formed in the form of microcrystals. Ti-(Fe)-oxide containing variable amounts of uranium is often found in uranium-ore black formations; this "titanate" (like rutile, anatase) is not yet more precisely diagnosed. Disperse mineral impurities can occur in uranium blacks: quartz, clay minerals, molybdenite, organic residues, etc., study of which was not within the scope of author's tasks.

Many years of black uranium ores study have shown that the terms "*uranium blacks*" or "*black*" ores are valid only as *morphological* characteristic of sooty black radioactive material.

It is possible that further studies of black material composition (its uranium-ore components) may lead to the discovery of additional mineral types of uranium, new for this material.

2.3.2.1 Uraninite

The main components of uranium blacks are U-oxides of various degrees of crystallinity (uranium pitch ore, pitchblende, nasturane). Numerous our microdiffraction studies confirm the crystallochemical belonging of natural uranium oxides to one single mineral species—uraninite, that was previously noted by many researchers (Kerr 1951; Sidorenko 1978, etc.). The subdivision of simple uranium oxides into *pitchblende*, *uranium pitch ore* and *uranium blacks* adopted in mineralogy reflects only their external, morphological character, but on diffraction they all represent one mineral with a cubic face-centered lattice of fluorite type; the spatial group (sp. gr.) *Fm3m*. Uraninite unit cell parameters decrease with oxidation degree increasing. By literature data there is genetic tendency: oxides with maximum parameter values limited to endogen ores, and with minimum—to the exogenous ones.

Uranium oxides with uraninite diffraction characteristics (a_0 = 5.42–5.38 to a_0 = 5.47–5.49) are most common in uranium blacks. The EM images (Fig. 2.7) show that uranium oxide particles are dense formations without defined outlines, with rounded or angular edges, and in some cases the spherical shape of aggregates is guessed. More often morphology of micron uraninite particles is the same type: lumpy particles, which are often aggregate of even smaller (submicron) particles. The diffraction of the different particles may be point, ring or completely absent (Fig. 2.8a).

The possibility of diffraction study of individual U-oxides mineral particles, with composition control of each of them, led to following conclusions. The finely dispersed nature of uranium oxide formations helps to rapid oxidation of U^4

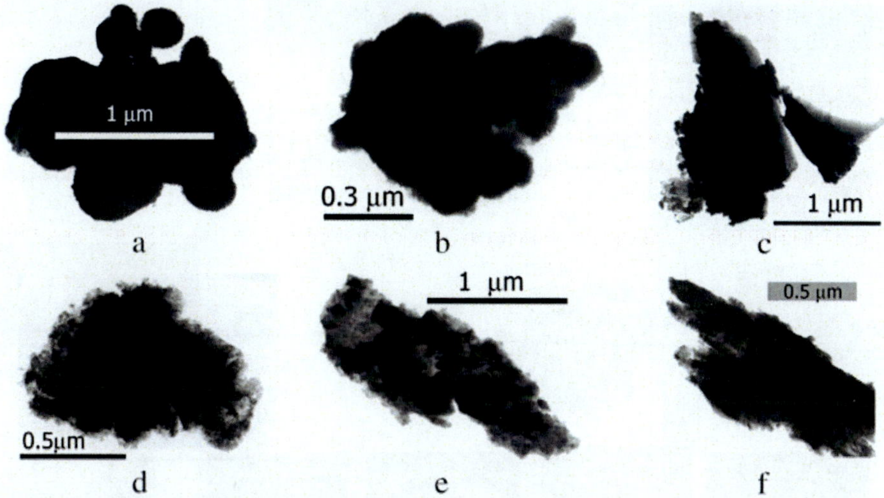

Fig. 2.7 Uraninite: micro-particles, TEM—image

$^{+} \rightarrow U^{6+}$, proportional to degree of material dispersion, that was noted earlier (Sidorenko 1978; Sidorenko et al. 1986; Dubinchuk et al. 1990). In one and the same sample, as U-oxides dispersity degree increases, the corresponding ring electronograms begin to dominate statistically over dotted ones during study. The oxidation process reduces the particles crystallinity degree, which is clearly visible in microdiffraction (MD) patterns. As structural perfection decrease (due to increased structural disturbances during oxidation), point diffraction reflexes disappear and become circular ones. Reflexes become wider, their quantity decreases, up to preservation of only one wide halo joining former reflexes 111 and 200 (d = 3,2–2,5). Ores with corresponding mineral substance gradually become X-ray amorphous and then electron-amorphous (Fig. 2.8a). Namely, degree of uraninite crystallization may be so small that it does not allow determining the mineral even with help of electronic diffraction. On the electronogram corresponding to such "non-structural" uraninite formations, at best, only weak intensity thickening in the first two uraninite rings area is fixed. Electron amorphous uraninite formations are quite common in natural samples; and their diagnostics on devices that are not equipped with microanalyzer is problematic. In this case of amorphous uranium-ore matter, the main diagnostic feature is elemental composition, determined directly in the electron microscope, and indirect feature is characteristic particle morphology.

In composition of poorly crystallized uraninite, with diffuse ring electron diffraction (ED) reflexes, Si is often present (Fig. 2.8a). Si impurity amount increases as crystallinity degree decreases. For sufficiently crystallized uraninite (point ED patterns), which are defined by mineralogists as nasturane, it is also possible Si presence in composition, but in much smaller amount; also possible to have insignificant Pb and Ca admixtures (Fig. 2.8c).

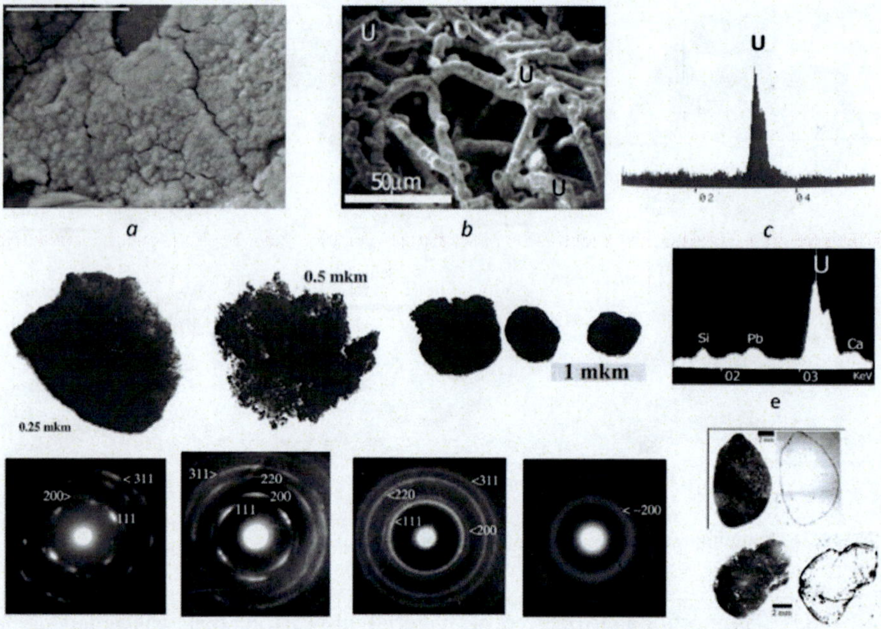

Fig. 2.8 Uraninite (U): (**a**) nasturane crust on sandstone particle (SEM—image); (**b**) biogenic forms of uranium-oxide (microalgae?); (**c**) EDS spectrum from nasturane crust; (**d**) series of nasturane particles (TEM-images) and corresponding microdiffraction patterns showing different crystallinity degrees; (**e**) EDS spectrum of an electron-amorphous nasturane particle (with the least perfect structure); (**f**) distribution of uranium in quartz pebbles (on left—thin section micrograph; on right—direct uranium distribution imaging on mylar film)

When black mineralization studying in oxidation zone one of uranium deposits, nasturane formation on uranophane has been fixed that indicates the processes of reduction subzone. Unusual uranium oxide biogenic formations in form of tangled threads have been recorded repeatedly (Fig. 2.8b). The "threads" diameter does not exceed 10 microns. Such formations resemble bacterial forms of Gallionella formed by Fe oxide, which were found repeatedly during EM study of Fe-Mn ores by F.V. Chukhrov and A.I. Gorshkov (Hypergenic… 1975). As judged by sizes exceeding usual bacterial formations, such forms are likely formed by micro-algae (according to Dr. Biologic Sci. N.V. Verkhovtseva, Moscow state University, oral report).

In exogenous ores, uranium oxides may be associated with carbonaceous organic material, but they are much more often associated with cementing material, or act as cement themselves. Such connection between uranium oxides and cementing material is very typical for hydrogenic uranium ores of sedimentary cover in deposits roll—and sandstone-type. Many years experience in studying black mineralization from various exogenous and epigenetic deposits allowed us to clarify some aspects of its genesis, namely: to detail the process of uranium oxides formation and to consider their deposition sequence.

Formation process of uranium ore substance in loose sandstone ores, where uranium oxides often act as cement, is presented as follows. The initial stage of formation of pitchblende ore cement (nasturane) is process of uranium sorption around grains of detrital material, primarily on quartz grains surface. In many spots of sandstone deposits where uranium content up to 0.n%, even the most effective modern mineralogical methods does not allow establish individual uranium minerals. Increased activity here is associated with the appearance of dark colored quartzes. And only f-radiography method shows that uranium tracks are confined to such dark quartz grains and fine pebbles, forming the thinnest rims around them (Fig. 2.8d). The degree of quartz darkening is directly proportional to the rims density. Upon reaching certain level of density in such rims, uranium oxides are fixed by ATEM method. Further uranium accumulation leads to release of individualized (separate) mineral phase in form of uraninite. In this way uranium oxides are formed in intergranular cement. Under favorable accumulation conditions, they play cementing material role. Organic residues in such ores are more often associated with precipitates coffinite.

2.3.2.2 Coffinite

This tetravalent uranium U^{4+} silicate is very widespread in composition of finely dispersed black ores, often as them significant part. Taking into account frequent Si-deficiency and (OH)-groups presence, coffinite crystal-chemical formula: $U^{4+}(SiO_4)_{1-x}(OH)_{4x}$ (Sidorenko 1978). It is tetragonal crystal system, sp. gr. *I41/amd*, isostructural with zircon and thorite. Coffinite is usually well crystallized as in exogenous deposits so in zones of secondary enrichment; and in hydrothermal deposits it is usually metamict, x-ray amorphous, that is usually due to more significant age of mineral.

In exogenous ores, it is often impossible to optically determine the signs that distinguish coffinite from ningyoite (U^{4+}-phosphate) and uraninite (when it is cryptocrystalline). In the majority of exogenous deposits, coffinites in thin rock section can be mistakenly diagnosed as ningyoites when diagnosed with coloration, reflective power and internal reflexes. It is likely that similar mistakes are also made in the study of hydrothermal ores, when only optical diagnostics of fine U-ore precipitates is carried out without the involvement of EM. Therefore, for diagnostics of X-ray amorphous hydrothermal coffinites it is necessary to use the whole complex of analytical EM methods in addition to traditional optical methods.

Many exogenous deposits are characterized by high degree of coffinite dispersion; its crystals size is 1–10 microns. In EM suspension preparations, coffinite rarely has characteristic forms—needle, spindle, oval (Fig. 2.9). More often, these are varies in thickness particles of lumpy morphology, which are aggregates of still smaller formations (Fig. 2.10). Formless single-crystal coffinite fragments and

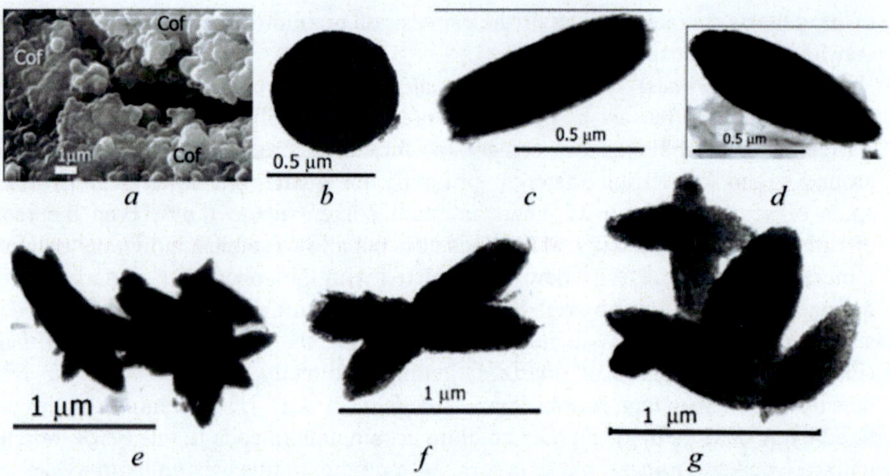

Fig. 2.9 Coffinite. SEM-image (**a**) and TEM-images (**b**, **c**, **d**) of individual microcrystals and their intergrowths (**e**, **f**, **g**—from Semizbay deposit)

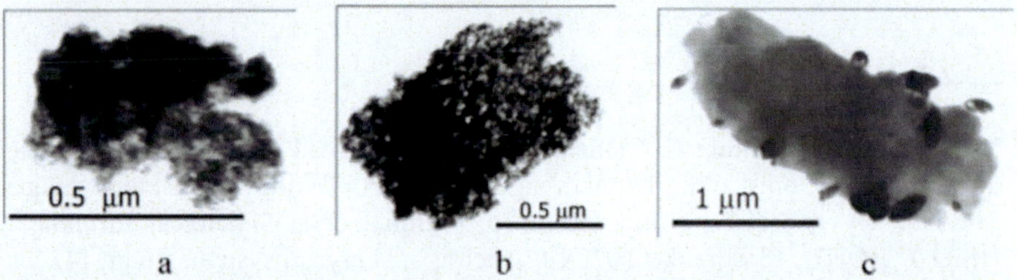

Fig. 2.10 Coffinite. TEM-images: (**a**, **b**) particles of monomineral aggregates, (**c**) coffinite microcrystals on layered silicate

particles in form of balls (Fig. 2.9b) are less common. Crystal sizes usually do not exceed several microns, sometimes there are twin of two microcrystals (Fig. 2.9c, e).

Coffinite characteristic feature is high crystallinity degree. Ring shaped microdiffraction patterns (MDP) of electronic diffraction (ED) are typical only for shapeless submicron aggregates of even finer particles (Fig. 2.11). In contrast to uraninite, coffinite has never been electron amorphous in all the exogenous ores studied. Coffinite MD—patterns are usually consisting of points (describing the single crystal) (Fig. 2.12). Even in the case of visually observable signs of leaching in EM image, when coffinite particles are "eaten" by dissolution, they often retain their point diffraction picture. Such ability to diffract as single crystal, when crystal particle continuity is broken, is interpreted by us as block isomorphism of mineral (according concept of Russian crystallographers N.V. Belov and V.A. Frank-Kamenetsky).

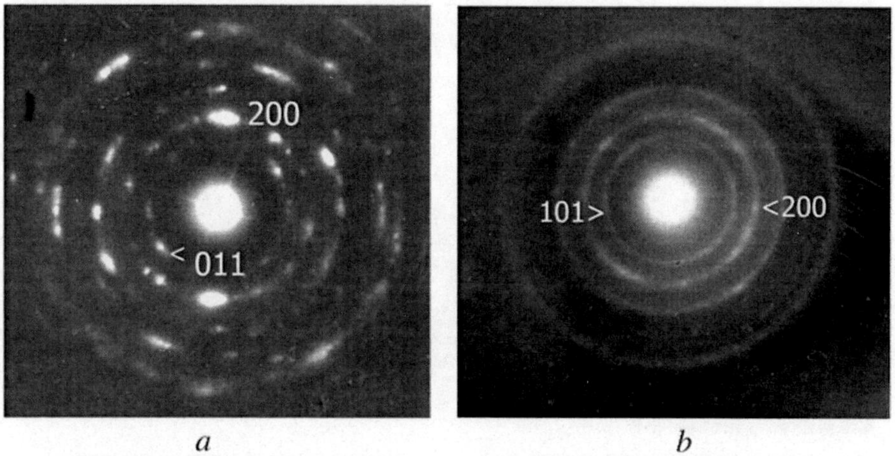

Fig. 2.11 Coffinite. Microdiffraction SAED patterns of the least crystallized particles: (**a**) arcuate-ring—and (**b**) ring-shaped

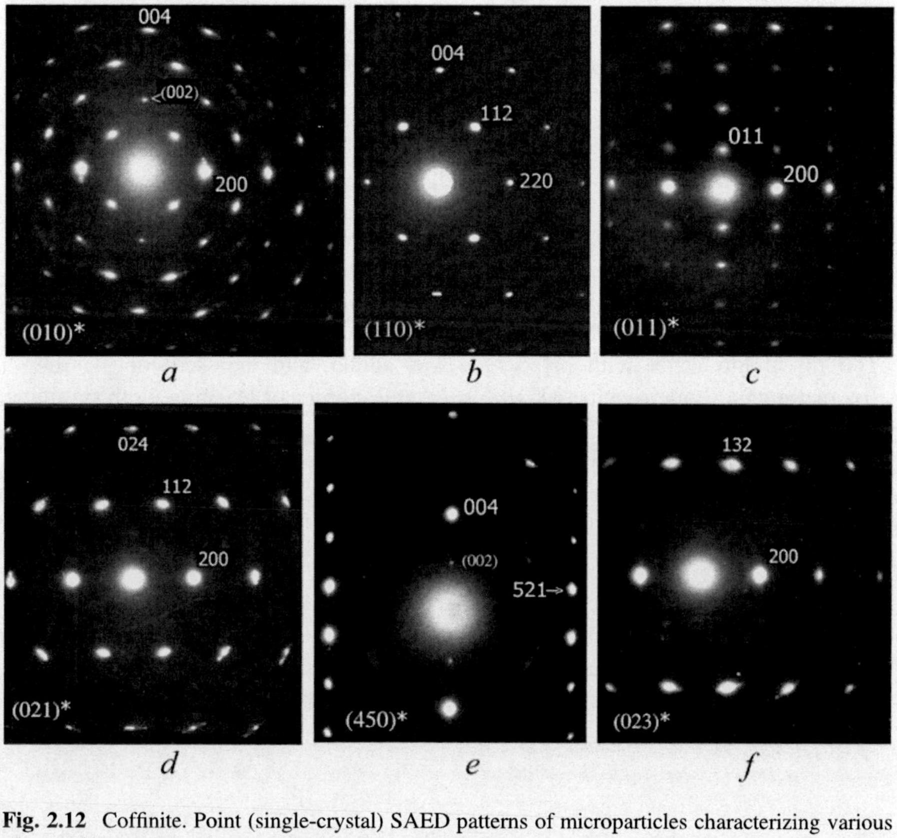

Fig. 2.12 Coffinite. Point (single-crystal) SAED patterns of microparticles characterizing various planes of the reciprocal lattice (indicated in parentheses)

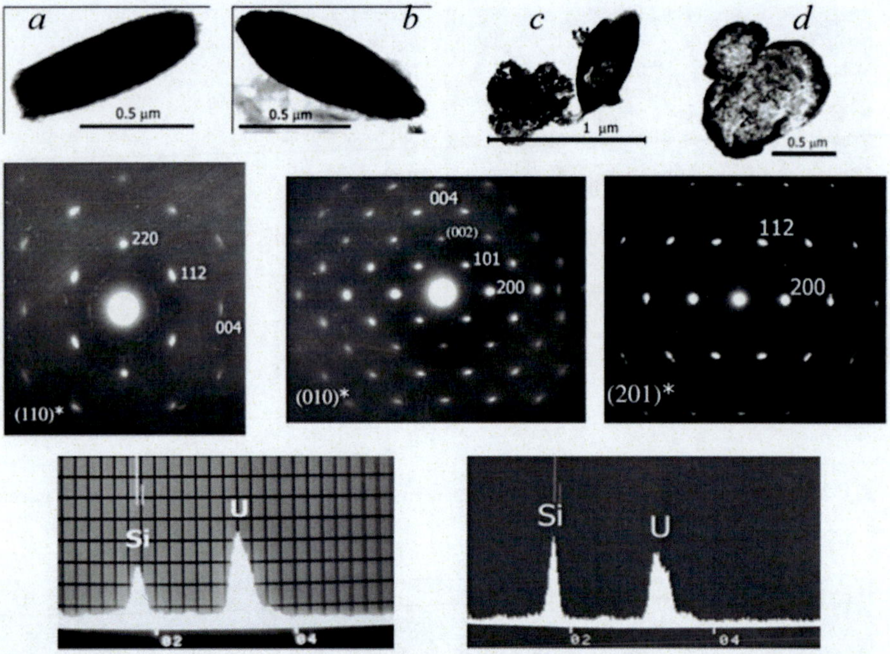

Fig. 2.13 Coffinite. Laboratory leaching of pure coffinite: (**a**) initial sample, (**b**) in 2 weeks, (**c**) in a month's time; (**d**) in 2 month's time. TEM—images of particles; corresponding SAEDs; EDS spectra before (left) and after 2 month (right) leaching

Laboratory experiments have confirmed the stability of coffinite structure, namely: its ability to keep the structure even with significant loss of uranium. Leaching was carried out in conditions close to industrial processing: unoxidized rock (20 g.), crushed to 200 meshes, was placed in flasks of 250 ml capacity, where 100 ml of tap water with pH $\approx$ 7.8 was added, and exposed for 2.5 months (by method in Shmariovich et al. 1982). After 2 months of leaching, with significant loss of uranium (peak U in the spectrum decreases to level of peak Si and below), the point diffraction of pure (without admixture) coffinite microcrystals fixes the keeping of structure without significant changes (Fig. 2.13). In the same conditions, coffinite which primarily contains impurities of P, V, Fe, As (Fig. 2.14), is completely leached and dissolved within a week.

Phosphorus-Containing Coffinite Often, in U—ore sites where clear positive correlation between uranium and phosphorus is observed, there is large number of ore samples with P-containing coffinite. It is very similar in shape to ningyoite. Microcrystals of coffinite in size from 0.5 to 2 μm have a needle shape, spindle-like

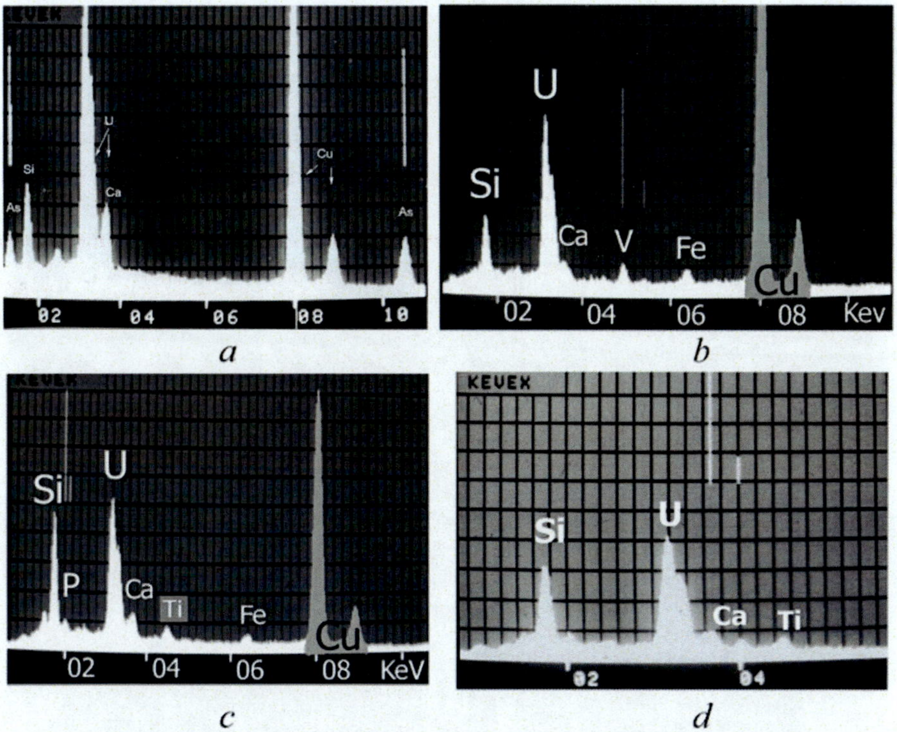

Fig. 2.14 Coffinite. Composition spectra of various coffinites with impurities As and Ca (**a**), Fe and V (**b**), P, Ca, Ti and Fe (**c**, **d**). Cu peaks from preparation

or oval (Fig. 2.15). Sometimes such coffinite is represented by massive spindle-shaped and oval particles with destroyed, partially dissolved ("corroded") middle. These particles, however, retain point reflexes of electron diffraction, indicating high degree of crystallinity. Such discontinuity of the mineral, its micro-blocky structure, indicates the natural process of ore substance processing (dissolution and replacement)—natural mineral "leaching".

In addition to U and Si cations typical for its composition, the particles of such coffinite also include P and, sometimes, Fe (Fig. 2.14c). Variability of P and Fe content from sample to sample in spindle-shaped coffinite microcrystals has allowed assuming that these elements are not included in the structure of coffinite or associated with it not strong enough. The experiment on leaching P and Fe from the particles of coffinite (Fig. 2.16), carried out directly on the EM preparation, showed that Fe (more easily) and P are almost completely removed, the U content decreases (Belova et al. 1980). For partially leached (dissolved) microcrystals, peaks integral intensities ratio of I_U/I_{Si} in the spectrum is the same as for impurity-free coffinites. Since electronograms from leached particles correspond to coffinite, it should be concluded that this leaching mode has no effect on coffinite structure (or this effect is insignificant). Elements P, Fe and partially U going into solution; the

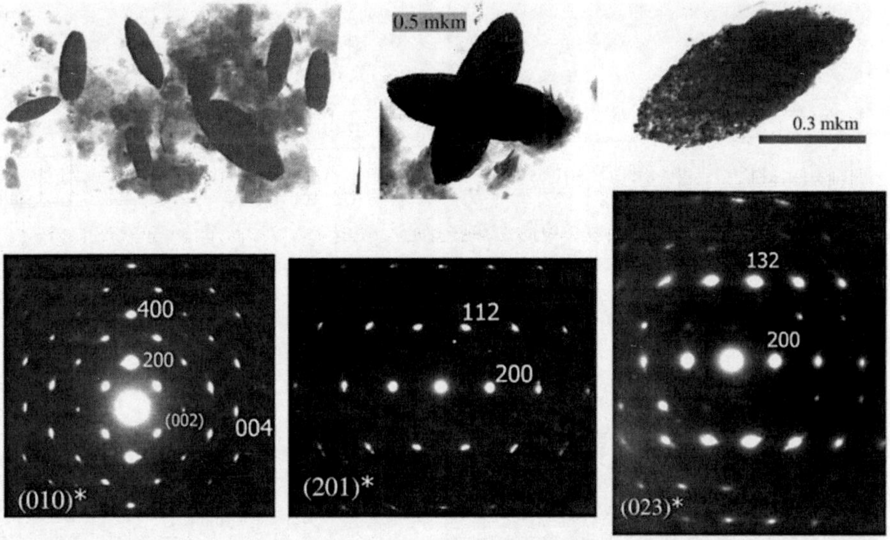

Fig. 2.15 P-containing coffinite. TEM image of particles in a suspension preparation (with layered silicate) and corresponding SAED-patterns characterizing the mineral

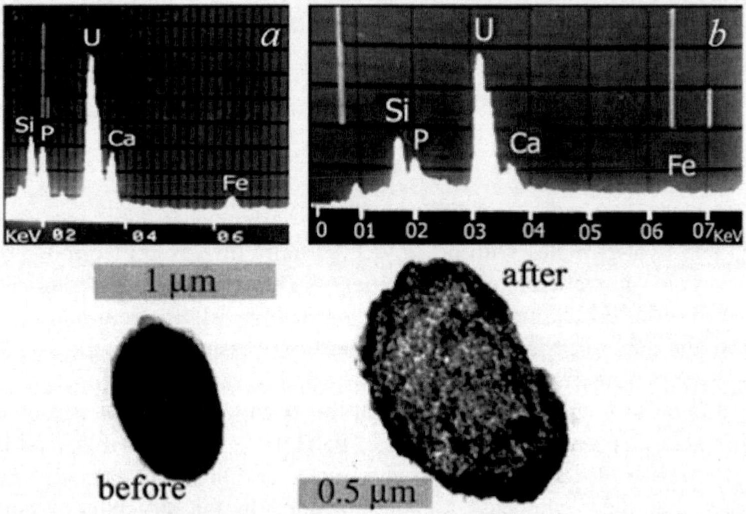

Fig. 2.16 Coffinite. Laboratory leaching of coffinite with impurities of P and Fe: TEM—particle image and ED spectra, (**a**) before leaching, (**b**) after leaching

nature of their relationship with coffinite is not yet clear. Taking into account the absence of any additional diffraction reflexes on electronograms (SAED patterns), it is possible to assume the presence of non-structural formations of phosphate composition (ningyoite?) associated with coffinite microcrystals, as proposed in our

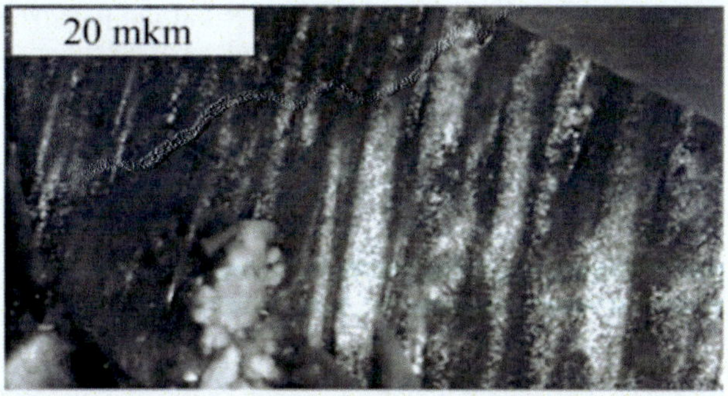

Fig. 2.17 Allocations of coffinite (white) in interlayers of carbonaceous matter (black). Khokhlovskoye, Yu.M. Dymkov's sample. BSE-image

publication on P-coffinite (Belova et al. 1980). However, it is more likely that so-called "block" isomorphism (Frank-Kamenetsky 1964) can be assumed here. Such character of isomorphism is consistent with the morphology of leached coffinites microcrystals and particles.

The data obtained substantiate the assumption about the possible formation of such (spindle-shaped) morphological type of coffinites as pseudomorphoses by ningyoite. Ningyoite-like form of coffinite microcrystals can be considered as hereditary, which is confirmed by the similarity of the coordinate faces parameters (010) of ningyoite ($a = 6.77–6.9$ Å; $c = 6.3–6.37$ Å) and coffinite ($a = 6.9–7.0$; $c = 6.21–6.31$ Å). Different content P (and Fe) in spindle-shaped coffinite particles in different samples, apparently, indicate some difference in the speed or duration of these elements removal by solutions during substitution process takes place, i.e. under conversion ningyoite into coffinite. Then it is reasonable to assume the possibility of reverse transformation of coffinites into ningyoite, which requires additional confirmation.

Thus, SAED study is basis for conclusions about coffinite behavior in leaching processes. Study of various and numerous samples allowed us to speak about mineralogical coffinite feature: in exogenous ores it is usually closely related to carbonaceous organic material (Fig. 2.17).

Note. As easily as impurity coffinite, finely dispersed uranium oxides and phosphates are leaching. Therefore, analysis of finely dispersed black U-ore material (that was subjected to oxidation) by traditional X-ray diffraction method alone may lead to erroneous diagnostics: against the background of X-ray amorphous uranium oxides, the real material composition of such ores will be defined as mono-coffinite. It is precisely these erroneous conclusions that were made in initially study of Chu-Sarysu deposits ores, the composition of which was later clarified as nasturane-coffinite one using ATEM methods.

2.3.2.3 Ningyoite

Tetravalent uranium phosphate ningyoite $CaU(PO_4)_2 \cdot 2H_2O$ has been considered for a long time as mineralogical rarity in contrast to uraninite and coffinite, that well known in uranium mineralogy and widely studied today. The mineral was discovered in 1957 (Muto et al. 1959) at the paleovalley uranium deposit Ningyo-Toge, Japan, and is named after the place of discovery. Since there have been no new discoveries of ningyoite for more than 20 years, the opinion about the uniqueness of this find in Japan was maintained among mineralogists. One of the reasons for this was probably the conclusions of Japanese mineralogists about the exceptional conditions of ningyoite formation (pH = 2), which are rarely found in nature (Muto et al. 1959; Muto 1962; Kajitani 1970). Long-term absence of new findings was considered by mineralogists as an indicator of extremely rare manifestations of ningyoite in nature. But the ATEM complex methods, first applied by us to study the fine-disperse material of uranium ores, has been shown wide distribution this Ca-U^{4+}-phosphate in nature. This mineral was found in U-ores of various deposits of the former Soviet Union (in hypergenesis zone) in different geographically remote regions; the mineralization mainly associated with stratal oxidation zones.

The second find in the world was our discovery of ningyoite on former Soviet Union territory in ore's composition of plast-infiltration deposit Sugraly, Kyzyl-Kum U-ore district, Uzbekistan (Belova et al. 1978a); in addition to ningyoite, uranium pitchblende (uraninite) broadly represented here in the samples.

Since the mid-1970s when the use of AEM methods began for mineralogical study, the little-known at that time ningyoitic mineralization was of particular interest of us. It was discovered by us (Belova et al. 1978a, b) two decades after the discoverers (Muto et al. 1959) in ores with a clear positive correlation of uranium and phosphorus (Sugraly, Kyzyl-Kum). It is such black ores that were more often chosen by us for further research in search of new finds of this mineral. As a result, ores of ningyoitic composition (Kosachinoe, Northern Kazakhstan; Momino, Haskovo, Navysen, Bulgaria) were studied in more detail (Doynikova et al. 2003).

Our subsequent ningyoite's numerous discoveries made it possible to obtain its crystal chemical data. Ningyoite was proved to belong to the mineral group of hexagonal aqueous phosphates—the rhabdophane group (Belova et al. 1985a, b). It was established isomorphic Fe entry to ningyoite composition (Belova et al. 1978a) and was showed wide cationic isomorphism manifestation (U, Ca, Th, REE) in its structure (Doinikova et al. 1993). Detailed description of the mineralogical characteristics of ningyoite is given next in Chap. 3 (Subsection 1).

It is necessary to underline characteristic feature of this mineral—extremely dispersed character of its precipitations: in all studied samples, size of crystals only in single cases exceed 1.5–2 microns (often n·0.1 micron); collomorphic formations are rare (sizes of up to several tens to hundreds of micron). Such a dimension of ningyoite also characteristic of Russian deposits in Khiagda Ore Field (Doynikova and Tarasov 2014). It is also necessary to note difficulty of ningyoite optical diagnostics in exogenous ores, and to note morphological similarity of its fine micron particles with coffinite—in form of needle and spindle-shaped microcrystals. For reliable diagnostics of these minerals (even at EM level) it is necessary to use SAED and EDS/VDS analysis of micron mineral particles.

This Ca-U^{4+}-phosphate is widely present in the hypergenesis zone. In uranium black composition it may participates as significant admixture or it can form monomineral formations. In all uranium ores in hypergenesis region, ningyoite is characterized by clear location on boundary of redox mediums and association with organic residues and sulphides.

As ningyoite is little known today mineral, all known at last time geological and mineralogical data on this U^{4+}-phosphate: its' findings history and geologic location, optical and crystallochemical data—are present more detail in special subdivision of next mineralogical division (Chap. 3).

2.3.2.4 Microbiological Factors in Ningyoite Ores Formation

Black U-ore location on mediums boundary coincides with propagation peculiarities of biochemical processes when uranium deposition onto reducing geochemical barrier. "The *most intensive development of biochemical redox processes is observed at the boundaries of heterogeneous media*", these "*manifestations of maximum microflora development*" induce "*...formation of ore mineralization in the hypergenesis zone especially in such places*" (Hydrogenic... 1980, p.71). In review article devoted to ningyoite formation geochemical conditions study (Kajitani 1970) the biogenic factor (vital activity of microorganisms) was recognized as determining factor in formation this Ca-U-phosphate (see next chapter). The decisive role played by *anaerobic* bacteria in uranium deposition from filtering solutions in ores of sandstone deposits has been repeatedly noted earlier (Hydrogenic... 1980; Dubinchuk et al. 1990; Kislyakov and Shchetochkin 2000).

In our studies, close association ningyoite with organic material is found in ores of practically monomineral phosphate deposits composition in Bulgaria, Momino and Navysen (Belova et al. 1978a, b, IGEM funds). These samples were loose dark brown powder, mixture of ningyoite and unexplored organic matter. This ore type samples always show the presence of framboidal pyrite (Fig. 2.18). Significant contribution of bacteria activity to formation of ningyoite mineralization can be traced back on Kosachinoye deposit, Kazakhstan (see above). The friable character of studied ningyoitic mineralization (micron crystallinity) and always close

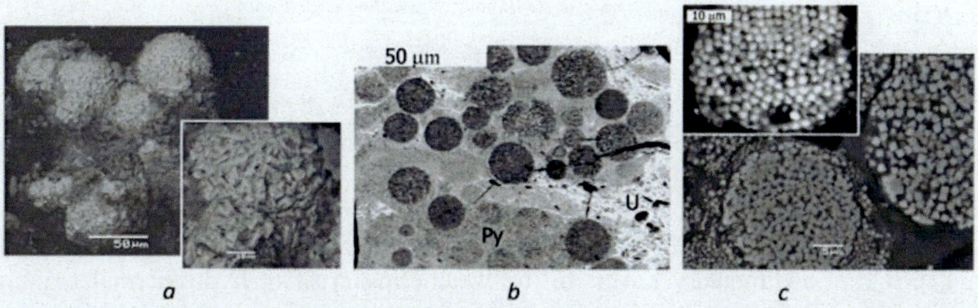

Fig. 2.18 Sulphide framboids: (**a**) marcasite intergrowth in ore sample (Khiagda deposit, Vitim), inset—enlarged fragment; (**b**) pyrite framboids (Py) associate with nasturane (U, white) in thin section of ore sample (Khokhlovskoye deposit, Trans Ural); (**c**) pyrite framboids in organic-mineral mass with U, thin section of ore sample (Vershinnoye deposit, Vitim); inset—pyrite framboid in powder ore sample (Bulgaria). BSE-images

association with organic matter allow taking opinion about decisive role of microbiological processes in mononingyoite ores formation. The leading role of such processes, which create reductive environment in near-surface sedimentary strata, is confirmed by the results of modern research on environmental mineralogy (see Chap. 6).

In recent decades, mineralogists studying uranium ores localized in hypergenesis region have increasingly focused on the important role of microbiological processes in ore formation. For example the publication of Japanese geochemists devoted to modern redox conditions study at uranium deposit Tono (Iwatsuki et al. 2003). This deposit, which ores are composed mainly of ningyoite, is located in uranium-ore region of Japan shared with Ningyo-Toge one. Redox buffer capacity of water-microbial system in and around the field has been studied to assess the long-term redox system stability. Tono U-deposit is located in the lower parts of Miocene sedimentary rocks series that overlapping granite as unconformity. The subsurface deposit waters contain sulphate and sulphide ions which are sensitive react on redox medium conditions. The microbes' abundance and viability in these waters, as well as isotopic ratios of sulphate sulfur, served as basis for the conclusion: microbial processes sulphate reduction and subsequent pyrite deposition, that involving organic matter in ore formation process,—are dominant redox reactions at the depth of uranium ore bodies. This conclusion is in good agreement with our numerous observations of pyrite framboids in black U-ores (Fig. 2.18, etc). Idiomorphic outlines and the undisturbed surface of pyrite microcrystals in such formations (its' have not affected by change or dissolution), allow to consider them as relatively young, formed not earlier than ningyoite. Calculations carried out by T. Iwatsuki and coauthors (Iwatsuki et al. 2004), using the content of sulfur (from sulphate rock) and the rate of sulphate dissolution, show that exclusively microbial sulfate reduction can maintain the necessary recovery mode for the conservation of uranium ores for several hundred thousand years (with unchanged hydrogeological conditions).

2.4 Conclusions

The above discussed features of uranium black mineralization, which forms industrial U—ore accumulations in hypergenic environment, allowed doing next conclusions.

It should be noted once again the similarity of mineral-forming processes both in the weathering crusts of hypogenic (hydrothermal) uranium deposits so in infiltration deposits of sedimentary cover. In the weathering crust of hydrothermal uranium deposits, the secondary enrichment zone (of commercial interest) is typical only for primary uranium ores enriched with sulphides. The formation of this zone (with regenerated uranium mineralization) is due by the obligatory presence of reducting factors (sulfides or organic material). Considered above secondary ores of Kosachinoye is example of untypical for hydrothermal deposits secondary enrichment zone formation, when functions of reducing factor performs an organic material, instead of sulfides.

Ores of infiltration sandstone type uranium deposits, in essence, represent a similar, widely manifested mineralization of reducing environment (zone); but it is more difficult to distinguish here the secondary enrichment zone boundaries. As at oxidation of sulfide-uranium ores, the mineralization scale here is depends on the presence of reducing factors in uranium-bearing layers. Such reducing agents are mainly various forms of organic matter—bacteria and other microorganisms connected with carbonaceous organic matter.

Disperse mineralization of tetravalent uranium—uranium blacks—are formed in reducing conditions at redox activity of weathering processes, they are shown practically in all deposits of uranium in hypergenesis region. Black uranium ores are typical for infiltration roll-type sandstone hosted deposits in sedimentary cover. Such ores are localized in sandstone strata in the form of cementing material of sandstone in the form of dark grey or black powder; that is usually represented by loose, slightly connected material. Uranium U^{+4}—mineralization is a bright "marking" sign that indicates the boundary of the reducing medium in the redox mineral zonality.

In essence, nature of sedimentary strata accumulation itself provides the existence of reducing factors, which serve as precipitating barrier at filtration of uranium-bearing oxygen-containing aqueous solutions. Therefore, the sedimentary strata in hypergenesis region can be considered as environment that is always favorable for the creation of black uranium-ore mineralization when presence source of water-soluble forms of uranium. Such source may be, for example, crystalline arrays that was opened by denudation and was eroding by surface waters. Such examples are well known in uranium deposits geology of sedimentary cover. The biogenic genesis of uranium blacks, with evaluating of microbiological processes contribution, will be considered further (Chap. 6).

Active development of environmental mineralogy and geomicrobiology in connection with immobilization of radionuclides in near-surface environment significantly helps forward study of exogenous U-ores formation.

References[1]

Belova LN (1975) Oxidation zones of hydrothermal uranium deposits. Nedra, Moscow, p 173. [in Russian]

[1] *These literary sources are kept in the funds of IGEM:*
Rybalov et al. IGEM funds, 1956
Gritsaenko et al. 1958 IGEM funds
Gritsaenko et al. IGEM funds, 1956
Geologic reports, IGEM funds, 1988, 1977

Belova LN (2000) Formation conditions of oxidation zones of uranium deposits and uranium mineral accumulations in the hypergenesis zone. Geol Ore Deposits 42(2):113–121. [in Russian]

Belova LN, Doinikova OA (2003a) Conditions of uranium minerals formation in oxidation zone of uranium deposits. Geol Ore Deposits 45(2):148–151. [in Russian]

Belova LN, Doinikova OA (2003b) Conditions of uranium deposits redox zonality formation in zone of hypergenesis. New in Earth Sciences. Abstr. Intern. Conf. Moscow. 2003. MGGRU [in Russian]

Belova LN, Gorshkov AI, Doynikova OA, Sivtsov AV (1991) Uranium black. In: development of V.I. Vernadsky's ideas in geological sciences. Nauka Publ, Moscow, pp 165–172. [in Russian]

Belova LN, Gorshkov AI, Ivanova OA (1978a) The first find of ningyoite in the USSR. Reports of the USSR Academy of Sciences. Dokl Akad Nauk SSSR 238(1):215–216. [in Russian]

Belova LN, Gorshkov AI, Ivanova OA (1978b) New data on ningyoite: Fe-containing ningyoite. Ibid 243(4):1022–1023. [in Russian]

Belova LN, Gorshkov AI, Ivanova OA, Sivtsov AV (1980) Nature of so-called P-bearing Coffinite. Ibid 255(2):428–430. [in Russian]

Belova LN, Gorshkov AI, Ivanova OA, Sivtsov AV (1985a) Ningyoite in the light of new experimental data. Proceedings of the USSR Academy of Sciences. Izvestiya AN SSSR Ser Geology 3:101–109. [in Russian]

Belova LN, Ivanova OA, Gorshkov AI, Sivtsov AV (1985b) Investigation of fine-dispersed uranium ores by complex of electron microscopy methods. In: Methods for studying uranium deposits in sedimentary and metamorphic strata. Nedra, Moscow, pp 158–175. [in Russian]

Boyle DR, Littlejohn AL, Roberts AC et al (1981) Ningyoite in uranium deposits of south Central British-Columbia, first north American occurrence. Canad Miner 19(4):325–331

Cai C, Li H, Qin M, Luo X, Wang F, Ou G (2007) Biogenic and petroleum-related ore-forming processes in Dongsheng uranium deposit, NW China. Ore Geol Rev 32:262–274

Doinikova OA (2007) Uranium deposits with a new phosphate type of blacks. Geol Ore Deposit 49 (1):80–86. https://doi.org/10.1134/S1075701507010047

Doinikova OA, Gorshkov AI, Belova LN, and others (1993) Questions on systematics of phosphates rhabdophane group. Zapiski VMO 3:79–88. [In Russian]

Doynikova OA (2005) Dispersed uranium minerals of hypergenesis reducing zone: mineralogy and crystallochemistry. Habilitation thesis. IGEM RAS, Moscow, p 277. [in Russian]

Doynikova OA (2012a) Mineralogy of uranium in reduction zone of hypergenesis (according to electron microscopy data). Fizmatlit, Moscow, 216 p.

Doynikova OA (2012b) Uranium mineralogy in hypergenesis reduction zone (according to electron microscopy data). Fizmatlit, Moscow, 216 p. [in Russian]

Doynikova OA, Belova LN, Gorshkov AI, Sivtsov AV (2002) The polymineral nature of uranium blacks. Uranium deposits: from their genesis to their environmental aspects (Proc. intern. Workshop. URANIUM 2002) Czech Geol. Survey, Prague, pp 45–48

Doynikova OA, Belova LN, Gorshkov AI, Sivtsov AV (2003) Uranium blacks: questions of genesis and mineral composition. Geology of Ore Deposit 45(6):514–530. [In Russian]

Doynikova OA, Tarasov NN (2014) Mokhov a.V. new Phosphatic type of uranium deposits in Russia. Dokl Earth Sci 457(Part 2):910–914. https://doi.org/10.1134/S1028334X14080030

Dubinchuk VT, Kochenov AV, Ruzhitsky VV, Meshchankina VI (1990) Forms of precipitate uranium mineralization of exogenous epigenetic ore-formation in sedimentary rocks according to electron microscopic studies. Lithology and exploitable minerals 3:65–72. [In Russian]

Dymkov Yu M (1973) Nature of uranium-pitchblende ore. Questions of genetic mineralogy. Atomizdat, Moscow, p 240. [in Russian]

Evseeva LS, Perelman AI (1962) Uranium geochemistry in hypergenesis zone. GosAtomizdat, Moscow. 239. [In Russian]

Frank-Kamenetsky VA (1964) Nature of structural admixtures in minerals. Nauka, L-d, p 263. [in Russian]

Gorshkov A.A. et al. (1988). IGEM funds. Annual report

Hydrogenic uranium deposits (1980) Ed. Perelman AI. Atomizdat, Moscow: 270 (in Russian)
Hypergenic iron oxides in geological processes. Moscow: Nauka. 1975—207. [in Russian]
Iwatsuki T, Arthur R, Ota K, Metcalfe R (2004) Solubility constraints on uranium concentrations in groundwaters of the Tono uranium deposit, Japan. Radiochim Acta 92:789–796
Iwatsuki T, Murakami Y, Naganuma T, Hama K (2003) Chiyukagaku (Geochemistry) 37:71–82
Kajitani K (1970) Econ Geol 65:470–480
Kashirtseva MF (1970) Methods for studying epigenetic changes in loose sedimentary rocks. Nedra, Moscow, p 152. [in Russian]
Kerr PF (1951) Natural black uranium powder. Science 114:91
Kislyakov YM, Shchetochkin VN (2000) Hydrogenic ore formation. Geoinformark, Moscow 608. (in Russian)
Lisitsin AK, Kuznetsova EG (1967) On the role of microorganisms in the formation of reducing geochemical barriers on the pinching-out of zones Plast limonitization. Izvestiya AN SSSR Ser geol (1) [in Russian]
Melkov VG, Pukhalsky LC (1957) Searches for uranium deposits. Gosgeoltehizdat, Moscow, p 220. [in Russian]
Min M, Xu H, Chen J, Fayek M (2005) Ore Geol Rev 26:198–206
Muto T (1962) The precipitation environment of ningyoite. Miner J 3(5–6):306–309
Muto T, Meyrowitz R, Pommer AM, Murano T (1959) Ningyoite – a new uranous phosphate mineral from Japan. Am Mineral 44(5–6):633–639
Perelman AI, Kasimov NS (1999) Landscape geochemistry. Astrea-2000, Moscow, p 768. (in Russian)
Rare-metal uranium ore formation in sedimentary rocks (1995) Collection of scientific papers. Science, Moscow, p 256. [in Russian]
Shmariovich EM, Zhiltsova IG, Pakulnis GV (1982) The role of pH changes in the formation of primary uranium-mica (uranite) mineralization. Bull USSR Acad Sci Ser geol (2):33–43. [in Russian]
Sidorenko GA (1978) Crystal chemistry of the uranium minerals. Atomizdat, Moscow, p 216. [in Russian]
Sidorenko GA, Gorobets BS, Dubinchuk VT (1986) Modern methods of mineralogical analysis of uranium ores. Energoatomizdat, Moscow, p 184. [in Russian]
Smirnov SS (1955) Oxidation zone of sulfide deposits. M.-L.: Publ. H. USSR Acad. Sci.: 1955—330 [in Russian]
Strakhov NM (1957) Stages of sedimentary rocks formation and tasks of their study /methods of studying sedimentary rocks, vol 1. Gosgeoltehizdat, Moscow, pp 7–28. [in Russian]
Uranium (2016) Resources, Production and Demand. OECD (2016) NEA/IAEA/ 7301
Vernadsky VI (1954) Selected works. v. 1. The briefs of geochemistry. USSR Acad. Sci. Publ, Moscow. (in Russian)
Vinichenko PV (2004) Theory of biogenic ore formation exemplified by uranium deposits. Sosnovgeologiya, Irkutsk. [in Russian]
Wülser P-A, Brugger J, Foden J, Pfeifer H-R (2011) The sandstone-hosted Beverley uranium deposit, Lake Frome Basin, South Australia: mineralogy, geochemistry, and a time-constrained model for its genesis. Econ Geol 106:835–867
Zhao L, Cai C, Jin R, Li J, Li H, Wei J, Guo H, Zhang B (2018) Mineralogical and geochemical evidence for biogenic and petroleum-related uranium mineralization in the Qianjiadian deposit, NE China. Ore Geol Rev 101:273–292. https://doi.org/10.1016/j.oregeorev.2018.07.025

Chapter 3
New Minerals Family: U^{4+}-Phosphates

In contrast to the widely known uranium $(UO_2)^{2+}$—phosphate—uranite or uranium mica—the phosphates of tetravalent uranium U^{4+} are practically unknown. By the mid-1970s, when ATEM dispersed uranium mineralization studies began, only one tetravalent uranium phosphate, ningyoite, was known among uranium minerals. Another mineral, lermontovite, which was discovered as uranium "silico-phosphate" in Caucasus in the 1950s, was considered doubtful. Insufficient study of the single find of lermontovite did not exclude the possible identity of this mineral to the later discovered ningyoite.

After the discovery of the first U^{4+} phosphate, ningyoite, in Japan at the end of the 50s, there were no other finds of this mineral or any new U^{4+} phosphates for a long time. Due to the possibility of elemental composition analyzing of micron crystals directly in the electron microscope, detailed study of material of the first find of lermontovite (V.G. Melkov's sample) by ATEM methods allowed to prove the individuality of this U^{4+} phosphate and to clarify its composition (Melkov et al. 1983). Further studies of finely dispersed exogenous U-ore mineralization composition have led to discovery of two new minerals—U^{4+} phosphates: vyacheslavite (Belova et al. 1983, 1984a, b) and urphoite (Belova et al. 1996).

Attempts at Systematics of the U^{4+}-phosphate mineral group and generalization of all available information were first undertaken in a brief review (Belova et al. 1998), where optical, chemical and structural minerals characteristics were summarized in a single table. This section content is the result of listed U^{4+}-phosphates detailed study that was considered in author's Dr. dissertation (Doynikova 2005) and then published as monograph (Doynikova 2012).

Four tetravalent uranium phosphates are known to date. One of them—ningyoite, $CaU(PO_4)_2 \cdot 2H_2O$—belongs to rhabdophane mineral group, and others are united for the present in single mineral group (Table 3.1), with common idealized formula U $(PO_4)OH \cdot nH_2O$, but with different structural characteristics. It is possible that in the future, when refining crystallochemical data at new finds, minerals of this group will be individualized, i.e. each of them can be representative of independent group. Consequently, in uranium minerals classification can be separated taxon higher than

O. A. Doynikova, *Uranous Mineralogy of Hypergene Reduction Region*, Springer Mineralogy, https://doi.org/10.1007/978-3-030-67183-9_3

Table 3.1 Tetravalent uranium U^{4+}-phosphates family (crystallochemical characteristics)

Minerals	Chemical formula	Structural data (Å) Syngony, sp.gr.	color, density ρ (g/cm^3)	Optic constants	Data source
Group of Rhabdophane $X{\cdot}(PO_4){\cdot}nH_2O$, X = Ce, U, Th, Ca, Fe					
Ningyoite	(U, Ca, Ce)$_2(PO_4)_2{\cdot}nH_2O$ $CaU(PO_4)_2$ $\cdot nH_2O$ n = 1–2	Orthorhombic, pseudohexagonal *hexagonal $P6_222$ $a = 6.86$, $c = 6.38$, Z = 3	Brown Brown, green in transmission light 4.74 (calc.)	$N_{cp} \approx 1.64$	Muto et al. (1959), Belova et al. (1985)
"Tristramite"	$(CaU^{+4}Fe^{+3})[(PO_4)(SO_4)(CO_3)]{\cdot}1.5H_2O$ Ca> > U> > Fe; PO_4> > SO_4	Hexagonal, $P6_222$ a = 6.913, c = 6.422 Z = 3	Greenish-yellow 3.8–4.2; 4.18 (calc.)	$N_o = 1.644$ $N_e = 1.664$	Atkin et al. (1983), Belova et al. 1987
Group of Lermontovite: $UPO_4OH{\cdot}nH_2O$					
Vyacheslavite	$(U_{1.061}Ca_{0.04})_{1.1}(PO_4)(OH)_{1.3}{\cdot}2.7H_2O$	*orthorhombic, *Cmcm, Cmc2₁, C2cm*; a = 6.96, b = 9.10, c = 12.38, Z = 8	Dark-green 4.6–5.2; 5.02 (calc.)	$N_g = 1.731–1.729$ $N_m = 1.729–1.726$ $N_p = 1.700$	Belova et al. (1984a, b)
Lermontovite	$(U_{0.94}Tl_{0.4}Ca_{0.02})_{1.0}(PO_4)(OH)_{1.2}{\cdot}0.4H_2O$ $(U^{4+}{}_{0.73}Ca_{0.005})_{0.74}(PO_4)$	*orthorhombic, *Ccca-?*, a = 9.74, b = 19.0, c = 10.1, Z = 8 Monoclinic $a = 10.00$, $b = 19.40$, $c = 8.55$; γ = 96	Green 4–4,5; 4,37 (calc.) — » —	$N_g = 1.724–1.726$ $N_m = 1.707$ $N_p = 1.686–1.690$ — » —	Melkov et al. (1983) Sidorenko et al. (1986)
Urphoite	$U_{0.85}Ca_{0.004}(PO_4)_{1.0}(OH)_{0.48}{\cdot}{\sim}0.3H_2O$	* orthorhombic, *Ibca*, a = 14.06, b = 13.22, c = 14.4; Z = 24	Emerald-green 4.29 (calc.)	$N_g \approx N_m \approx 1.734$ $N_p \approx 1.707–1.708$	Belova et al. (1996)

Note: * by SAED data

mineral group, taxon of U^{4+}-phosphates. It should be smaller taxon than mineral class, which unites all phosphates in common mineral taxonomy. Obviously, the most suitable name for that taxon unifying U^{4+}-phosphates would be term "*family*", which has a wide application. According to famous Russian crystallographer, maker of new minerals classification principles G.B. Bokiy, "this taxon combines mineral species by their similarity in composition ... regardless of minerals structure" (Bokiy 1997). Therefore, it is legitimate to speak about the discovery of new minerals *family*: tetravalent uranium phosphates (U^{4+}-phosphates).

Subsection devoted to ningyoite is significantly more than all others in chapter; this disparity is fully reflects prevalence degree (currently known) and knowledge degree on each mineral. Numerous ningyoite finds allowed us not only to determine its formation conditions in various geological localities, but also to clarify classification within mineral group to which it belongs—in Rhabdophane group (see Chap. 4). For other U^{4+}-phosphates, the data given here are essentially description of their single or only finds. Crystal-chemical data of all these minerals are limited by the amount of information available to ATEM methods when studying micron objects.

Tetravalent-uranium phosphates are found mainly in conditions where reducing stage of hypergenous redox processes is manifestation. That is typical as for hydrothermal uranium deposits oxidation zones and so for hydrogenic deposits sandstone-type (Belova and Doinikova 2003a, b; Doynikova et al. 2003). So-called blacks' mineralization is the bright "marking" sign of reduction zone in redox mineral zonality (see Chap. 2). Its study led to discovery of new U^{4+}-phosphates among blacks' uranium formations, such them association is very characteristic.

3.1 Ningyoite: $CaU^{4+}(PO_4)_2 \cdot nH_2O$—of Rhabdophane Group Mineral

A Brief History of Ningyoite Finds As noted above, ningyoite was discovered in 1957 (Muto et al. 1959) at the paleochannel uranium deposit Ningyo-Toge, Japan; idealized mineral formula $CaU(PO_4)_2$-$2H_2O$. The second world find was our discovery of ningyoite in the exogenous Sugraly uranium deposit ores, Kyzyl-Kum ore district, Uzbekistan (Belova et al. 1978a). Our second discovery of ningyoite in the weathering crust of hydrothermal deposit Kosachinoe in Northern Kazakhstan (Belova et al. 1978b) revealed its new, Fe-containing variety (Belova et al. 1978b).

Thanks to the applied set of ATEM methods for studying uranium matter of sandstone type deposits in sedimentary cover, ningyoite was discovered by us in 1976–1977 almost simultaneously in Soviet Union and Bulgaria (Belova et al. 1978a, b, 1979). In Bulgaria uranium ores, ningyoite is particularly widespread, and in number of deposits it is represented exclusively by Fe-containing variety.

From mid-1970s the wide AEM-using began for mineralogical study of fine dispersed material in different ores. Practically unknown in those years U^{4+}

phosphate, which we found in ores with clear positive correlation of U and P, attracted our particular interest. It is such black ores were more often chosen by us for further research in searching for new that mineral's finds. As a result, ningyoite's mineralization in deposits of Kazakhstan (Kosachinoye) and Bulgaria (Momino, Haskovo and Navysen) was studied in detail (Doynikova et al. 2003). When mononingyoite composition of black U-ores was established in Navysen, taking into account their similarity to Japan ningyoite ores, we first declared discovery of new type of black uranium ores with phosphate mineralization (Belova et al. 1986, IGEM funds). This was confirmed by subsequently discovery different infiltration sandstone type uranium deposits with predominantly ningyoite ores.

After Russian finds, data on ningyoite finds were published in uranium deposits of Northern Bohemia, Czechoslovakia (Scharm and Hofreiter 1978; Scharm et al. 1980) and in Lower Silesia, Poland (Kucha and Weiczorek 1980). Later deposits with predominantly ningyoite ores was find at hydrogenic uranium deposits in British Columbia, Canada (Blizzard, Tayi) (Boyle et al. 1981). Then, under the name of tristramite (see below), ningyoite was discovered in old mine dumps in Great Britain (Atkin et al. 1983; Belova et al. 1987). Based on ATEM methods, ningyoite was detected in several ore samples from Siberia, near Vitim River, at Khiagda field, Buryatia (Ryzhov et al. 1985). Detailed studies later have shown this mineral to be predominant uranium ore here (see Chap. 5).

Ningyoite of hydrothermal genesis was found in ores of five-metal (U-Bi-Co-Ni-Ag) formation of classic vein Deposit of Czech massif Horní Slavkov (Dymkov et al. 1986); in ore veins of uranium deposits Jáchymov and Rožna, Czech Republic.

Series of publications is devoted to results of ningyoites study from infiltration deposits in Northern Bohemia ore district, Czech Republic, by scanning analytical EM (Scharm 1993; Scharm et al. 1994; Scharmova et al. 1993; Scharmova and Scharm 1994).

On former Soviet Union territory until to 2000s, there were met only separate finds of ningyoite, which were made in various deposits associated with weathering crust (Kosachinoye, Kazakhstan), with secondary enrichment zones (Zaozernoye, Northern Kazakhstan), with plast—(Sugraly, Uzbekistan) and ground—(Khiagda, Northern Transbaikalia, Buryatia) oxidation zones. According to L.N. Belova (oral report), it is quite possible that in a number of other deposits ningyoite would have been found in higher concentrations if it were possible to use EM in U-ore samples studying.

In recent years, when geologists IGEM study Paleozoic uranium deposits on Vitim plateau in Siberia, it was proved ningyoite composition of their ores by using analytical EM methods for mineralogical investigations. Thus was revealed here new for Russia type of phosphate-uranium black's ores (Doynikova et al. 2014). Results of these new studies are presented below in Chap. 5.

3.1.1 Morphological and Optical Characteristics

Formation Morphology Ningyoite discoverers, with help of light optics, were found this mineral in form of tiny needle crystals covering pyrite or apatite, or filling cavities and cracks in pebbles. A characteristic feature of ningyoite is highly dispersed nature of its formations. In hypergenic conditions, ningyoite is manifested in the composition of uranium blacks. The monomineral formations of this Ca-U-phosphate are typical uranium black ones: it is loose powder of brownish black or dark brown (in mix with organic matter) color. In hydrothermal ores, ningyoite is manifested in form of granular masses and nodular crust, which encrusting nasturane or rammelsbergite (Dymkov et al. 1986).

Crystal Size and Shape: According to Electron Microscopy In all samples we studied, the size of ningyoite crystals was usually less than 1 μm, and maximum value did not exceed 1.5 μm. The largest columnar crystals established in ores of tStráž block, North Bohemia, Czech Republic (Scharm et al. 1980; Scharmova and Scharm 1994), reach 10–40 microns; the finds crystals up to 100 microns is noted here as an exceptional rarity.

According to our data, ningyoite microcrystals are of needle-like and spindle-shaped habitus (Figs. 3.1 and 3.2), which is confirmed by other researchers data (Muto et al. 1959; Scharm et al. 1980; Boyle et al. 1981; Atkin et al. 1983; Scharmova and Scharm 1994). The habitus of larger microcrystals is columnar (Fig. 3.3a, b) or narrowly columnar (Scharmova and Scharm 1994). In the Czech Republic, colloform formations of ningyoite with uraninite were found (Fig. 3.3b). Ningyoitic crystals are rhombic or hexagonal in cross-section (Figs. 3.3c and 3.4) and have fibrous-like appearance in nodular crusts and micro-spherulites (Dymkov 1985; Dymkov et al. 1986).

For the Fe-containing ningyoite variety, elongated (along ***c*** axis) microcrystals with clearly visible faces and faceted vertices are characteristic, what distinguishes them from ironless varieties of this mineral of needle-spindle-shaped. Fe-ningyoite

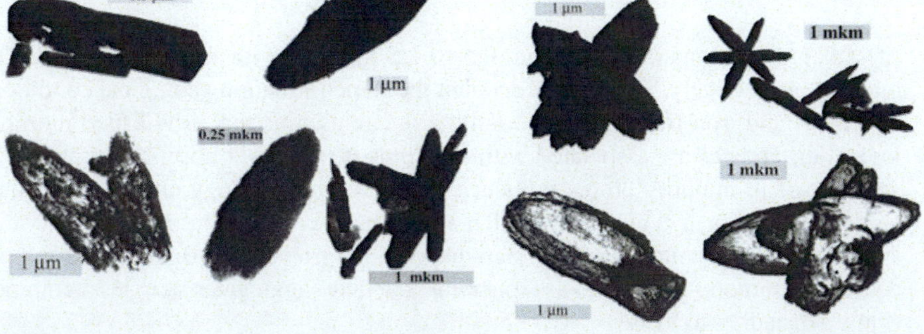

Fig. 3.1 Ningyoite. Individual microcrystals and their intergrowths; last two—carbon replicas. TEM-image

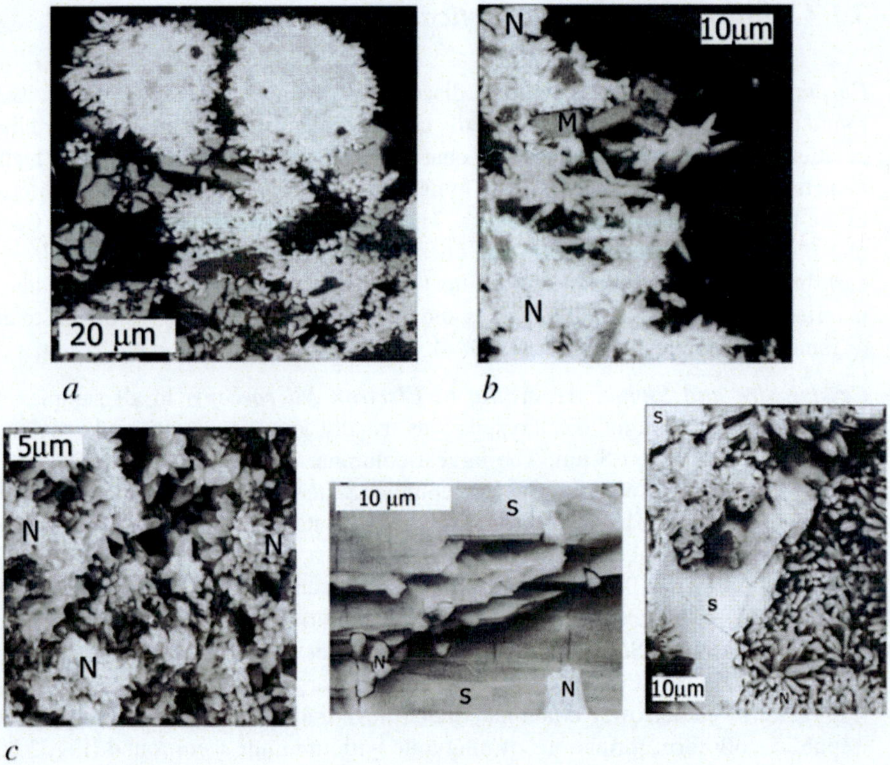

Fig. 3.2 (**a**) Uranium ore from Ningyo-Toge, Japan SEM (BSE)—image, polished section; sample from Ontario Museum, Canada; from Boyle et al. (1981). Spherolitic aggregates of ningyoite spindle-shaped crystals (white) in matrix of clastic quartz (gray) and carbonaceous material (black). Inside the ningyoite aggregates—pyrite (gray). (**b**) Uranium ore from Tyee, Canada SEM (BSE)—image, polished section; from Boyle et al. (1981). Spindle-shaped ningyoite (N) crystals and their star-like intergrowths are among quartz (Q), marcasite (M) and carbon material (black). (**c**) Uranium ore from Blizzard, Canada SEM (BSE)—image, polished section; from Boyle et al. (1981). Spindle-shaped ningyoite (N) crystals forming rosettes (on left), which surrounding the saléeite (S) and growing along cleavage cracks in it (two right photos)

crystals are also small: length from 0.5 to 1.5 microns with width from 0.1 to 0.3 microns, respectively. Taking into account the same formation geological conditions of ferrous and iron-free mineral varieties, we can assume that established morphological differences are associated with Fe atoms presence (or absence) in ningyoite structure. The impurity atoms influence on crystal morphology change was noted earlier (Petrovskaya et al. 1976), which allows us to consider our assumption valid.

For both ningyoite varieties, star-shaped intergrowths of 0.5÷2 microns are equally common, and same-as spherical star-like aggregates are less common (Figs. 3.2 and 3.3a, b).

In the hydrothermal veins of Horní Slavkov (Dymkov et al. 1986) the main mass of ningyoite is represented by xenomorphic formations (as filling cavities and

Fig. 3.3 (**a**) Ningyoite, Strazh, N Bohemia, Czech Republic SEM (BSE)—image, polished sections; from Scharmova and Scharm (1994). **a** Rosette-like aggregate of idiomorphic ningyoite crystals, dark zones and gray areas in some ningyoite crystals—pyrite (P); **b** Ornamental texture composed of ningyoite (light) and pyrite (gray). (**b**) Ningyoite, Strazh, N Bohemia, Czech Republic SEM (BSE)—image; from Scharmova and Scharm (1994). **a** Ningyoite (N) columnar crystals accreting with vyacheslavite (V); **b** intergrowths of ningyoite crystals; **c** weakly differentiated mineral substance of gel-like origin, its main components are similar in composition to uraninite (white, U), ningyoite (gray, N) and baddeleyite (dark gray). (**a**) BSE + SE-image; (**b**, **c**) polished sections, BSE-image. (**c**) Ningyoite, Strazh, N Bohemia, Czech Republic SEM (BSE)—image, polished sections; from Scharmova and Scharm (1994). Examples of ningyoite replacement: (**a**) with uranium oxide (white in ningyoite aggregates center); (**b**) substitution of ningyoite (white) with pyrite (gray); (**c**) ningyoite (gray) substitutes REE phosphate chemically similar to rhabdophane/monazite (light gray)

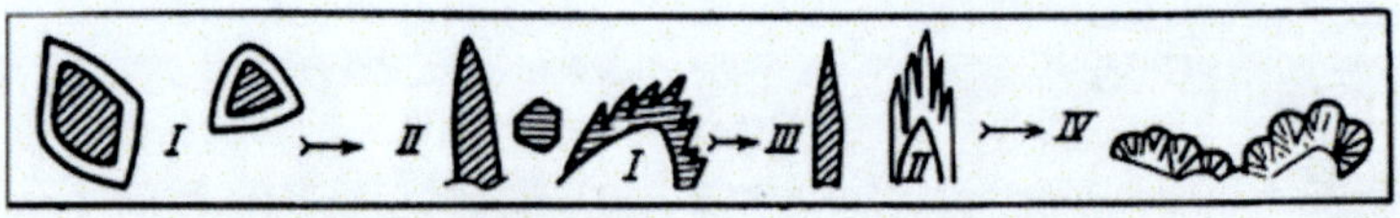

Fig. 3.4 Evolution of ningyoite crystals forms in time (according to Yu. M. Dymkov 1985): gradual elongation of crystals from almost isometric (plane-curved) to needle-shaped with a hexagonal section (I–II); tendency to elongate shapes with plane parallel faces (I–III); cleavage at the top of crystals (II); formation of spherulite crusts (II–IV). Micron crystal sizes keep stably

cementation of other minerals); less often ningyoite forms individual crystals up to 0.5–0.7 microns in length, crystal intergrowth and overgrowing crusting. Yu.M. Dymkov (1985) supposed evolution of ningyoite crystalline forms in time, during growth (Fig. 3.4).

Optical Characteristics Under optical microscope, the color of ningyoite is greenish-brown, green. In transmission, the mineral is brownish-green or brown. Pleochroism is weak, darker by *Ng*. The average values of refraction index $n = 1.64$ (Muto et al. 1959); $n = 1.60–1.70$ (Boyle et al. 1981). Positive elongation, extinction is parallel to elongation. No fluorescence. It is necessary to note the complexity of ningyoite optical diagnostics in exogenous ores and so morphological similarity fine micron formations of ningyoite and coffinite (of needle and spindle-shaped microcrystals).

Under ore microscope, in reflected light, ningyoite reflection is lower than that of coffinite (extremely low). When observed in dark field, ningyoite from hydrothermal veins was brown, but Y.M. Dymkov notes: "the primary color of ningyoite should be . . . in green tones"—similar to other phosphates of tetravalent uranium. "In reflected light among uranium minerals it is almost black, but among sulfides ningyoite crystals are transparent. . . ", "the mineragraphy study of ningyoite does not exclude possibility of erroneous identify. . ." (Dymkov et al. 1986).

In exogenous ores, it is often not possible to identify the signs that distinguish ningyoites from coffinites. When studying thin rock section, ningyoite you may mistakenly diagnose as coffinite or uraninite (in its cryptocrystalline formations) in the ores of most exogenous deposits, if diagnosis by color, reflectivity and by internal reflexes is carried out.

Therefore, in addition to traditional optical methods, it is necessary to use a full range of EM methods for diagnostics. Probably, similar mistakes are also made in the study of hydrothermal ores, when only optical diagnostics of fine ore formations is carried out without the involvement of EM. Moreover, elemental composition analysis of these minerals must be combined with SAED for trusting diagnosis.

3.1.2 Crystallochemical Characteristics

3.1.2.1 Diffraction Data

Ningyoite' discoverers determined its elementary cell parameters by X-ray powder photo method and obtained following results: orthorhombic syngony, $\boldsymbol{a} = 6.78 \pm 0.03$, $\boldsymbol{b} = 12.10 \pm 0.05$, $\boldsymbol{c} = 6.38 \pm 0.03$ Å; spatial group P222 (D_2^1), Z = 3 (Muto et al. 1959). It has been established the similarity debaegramms of natural ningyoite and its synthesized analogue to debaegramm of rhabdophane—hexagonal cerium phosphate.

By means of *powder X-ray investigation* the author studied two ningyoite samples: iron-free and Fe-containing. The basis for the X-ray patterns calculation, for the indexing of reflections, was taken from the microdiffraction SAED data (see methodical section I): cell parameters and relative reflex intensities, taking into account multiplicity. In samples by XRD studied, the predominant mineral substance part is represented by ningyoite: the first one is mixture of ningyoite and unstudied organics; the second one contains dispersed mineral impurities. To refine parameters of both samples, Si-standard was used at exposition. The high background level on Fe-ningyoite debaegramm (Fig. 3.5) and broadening of all ref-lexes were caused, probably, by high material dispersion; that caused low accuracy of determination of interplanar distances *d/n* (not higher ±0.01 Å; XRD cameras—86.0 and 114.6 mm; Fe-radiation). Diffractogram was obtained for significantly enriched sample of iron-free ningyoite with largest (~1.5 microns) ningyoite crystals (URS-50, Cu-radiation) (Fig. 3.6). This sample is represented by brownish-black powder of ningyoite and plant organics mixture. For both ningyoite varieties—iron-free and Fe-containing—the parameter values do not differ within the method accuracy (Fig. 3.7).

X-ray ningyoites study results with indication in rhombic and hexagonal syngony are given in the table, where all previously published results of X-ray ningyoite studies are also presented for comparison (Table 3.2). Parameters of hexagonal ningyoite cell refined by least-square method: $a = b = 6.87$ Å; $c = 6.40$ Å.

Parametric ningyoites characteristics were studied using SAED method. In was made comparative study of our first finds of this mineral and the sample from Ningyo-Toge, which was kindly provided to L.N. Belova by Dr. T. Watanabe. It was established that the rhombic structure of the mineral is characterized not primitive elementary cell, as determined by Japanese researchers, but is C-centered one. Reflex extinctions correspond to one of spatial groups: C222, Cmm2, Cm2m, and Cmmm. Microcrystal elongating along the axis $\boldsymbol{c}$. The elementary cell

Fig. 3.5 Debaegramm of Fe-ningyoite (powder sample, Navysen, Bulgaria; X-ray camera-86 mm; Fe-radiation)

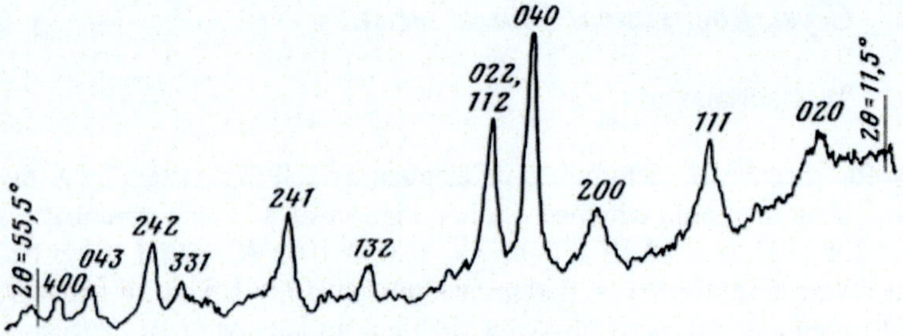

Fig. 3.6 Ningyoite. XR-diffractogram of mononingyoite sample (dark-brown powder of ningyoite-organic mixture; Momino, Bulgaria; Cu-radiation)

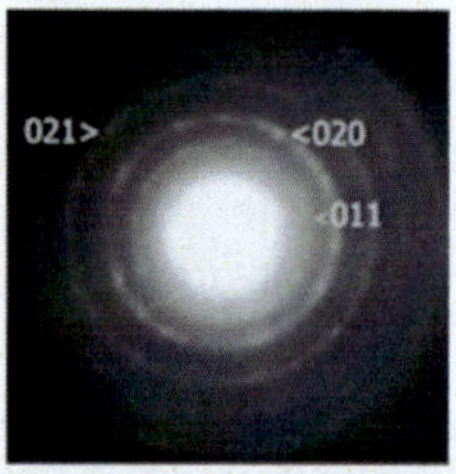

Fig. 3.7 Ningyoite: ring SAED-pattern

parameters calculated from the SAED patterns did not practically differ from those previously established by Japanese researchers. Within accuracy limits of method, no differences in parameters of two ningyoite varieties were found: ordinary iron-free and Fe-containing. The data on C-centered rhombic cell of ningyoite, given in our articles (Belova et al. 1978a, b), were later confirmed by researchers from British Columbia, Canada (Boyle et al. 1981).

Later, with careful SAED study of samples in which ningyoite is the only uranium mineral and is represented by larger microcrystals (1.5÷2 microns), the elementary cell parameters were precised; it was proved that true symmetry of ningyoite structure is hexagonal one (see Chap. 1).

The previously selected, C-centered cell with orthogonal crystallographic axes is actually orthohexagonal (Belova et al. 1985). These data were confirmed by parallel study of ningyoite sample from Japan as reference standards. Parameters of ningyoite hexagonal cell: $\boldsymbol{a} = \boldsymbol{b} = 6.86$; $\boldsymbol{c} = 6.38$ Å (±0.03). Calculated density $D_{calc} = 4.74$.

In proving hexagonal symmetry of ningyoite, an important role was played by equipping the transmission electron microscope by goniometer with large tilt range (120°), which allowed to consistently fix identical diffraction SAED patterns every 60° when rotating crystals around elongation axis (see Chap. 1). The specific morphology of microcrystals—elongation along by c-axis—was helpful to accurate orientation of obtained SAED patterns (crystal lattice cross-sections).

Table 3.2. Ningyoit X-ray powder results

Muto et al. 1959			Scharm and Hofreitr 1978		Scharm et al. 1980		Boyle et al. 1981			Belova et al. 1980, 1985		
I	*d*	*hkl* Orthorhomb	*I*	*d*	*I*	*d*	*I*	*d*	*hkl* Orthorhomb	*I*	*d*	*hkl* Hexagonal
5	12.0	010										
5	6.8	100										
20	6.35	001			20	6.31				16	6.38	001
40br	5.99	020	40	5.98	65	5.98	5	6.02	020	25 br	5.95	100
20	5.65	011			30	5.69						
20	4.49	120			20	4.53						
60	4.33	111	40	4.36	55	4.37	9	4.35	111	40	4.33	011
5	4.02	030										
20	3.45	130			50	3.44	6[a]	3.44	130			
60br	3.38	200	40	3.40						27	3.38	110
5	3.19	002								5	3.20	002
100	3.02	040	100	2.99	100	2.99	10	3.00	040	100	3.00	020, 111
80	2.81	022, 112	50	2.819	55	2.818	9	2.81	022, 112	70	2.81	012
20	2.73	041					1[a]	2.71	041			
20	2.69	221										
40	2.35	132	20	2.342	20	2.343	1	2.35	132	24br	2.48	021
5	2.33	202					3[a]	2.32	202	12	2.34	112
5	2.26	240								10	2.24	120
5	2.22	310										
80	2.13	241	50	2.130	90	2.129	8	2.13	003, 151	40	2.13	003
5	2.01	060					4[a]	2.01	113, 060			
5	1.93	061										
20	1.903	331			40	1.904				9br	1.90	031
40	1.845	242	30	1.846	30	1.842	5	1.843	152	33	1.84	122

(continued)

Table 3.2. (continued)

Muto et al. 1959			Scharm and Hofreitr 1978		Scharm et al. 1980		Boyle et al. 1981			Belova et al. 1980, 1985		
I	*d*	*hkl* Orthorhomb	*I*	*d*	*I*	*d*	*I*	*d*	*hkl* Orthorhomb	*I*	*d*	*hkl* Hexagonal
40	1.740	043					1	1.735	043	16	1.74	023
5	1.728	223					1/2	1.719	260			
40	1.692	400			15	1.691	4	1.691	062, 400	14	1.69	220, 032
20	1.681	332					1/2	1.674	332			
20	1.653	350					1	1.657	170, 261	11	1.65	130, 221
5	1.634	420					3[a]	1.594	004	5	1.59	004
20	1.547							2	1,536	313		
							1/2	1.518	262			
5	1.505						1	1.501	080			
							1	1.475	172			
5	1.451						1/2	1.450	422			
							1	1.432	441			
							1	1.363	370			
20	1.343						1/2	1.342	281			
20	1.307						2	1.304	154			
5	1.262											
5	1.247											

Note: [a]Lines reinforced with marcasite admixture; *br* broad

At present, the task of this U-Ca-phosphate structure determination using data of single crystal electron diffraction—by point reflexes intensity at SAED patterns—seems promising.

It is clear, however, that we can talk only about the calculation of preliminary reference model of ningyoite structure.

It is clear, however, that we can only talk about the calculation of the preliminary reference model of the structure of ningyoite on the basis of already determined structure of ordered low-temperature synthetic phase $CaU(PO_4)_2$ (Dusausoy et al. 1996). Considering that accuracy of d/n interplanar distances determining in electron diffraction is one order less than in X-ray radiography, it is obvious that to solve this problem cannot do without involvement of powder X-ray methods. Recently, the method of precession electron diffraction (PED), which is used for the structural study of nanomaterials, seems promising for the study of hexagonal aqueous phosphates.

3.1.2.2 New Structural Data on Ningyoite

In order to detail the structural features, the samples of natural ningyoite and natural rhabdophane, the same those studied by X-ray radiography, were studied by infrared spectroscopy (unpublished data of N.V. Chukanov). IR spectra of samples, pressed into the KBr tablets, were recorded on the Specord 75IR spectrophotometer. For comparison, IR spectra of both minerals—ningyoite and rhabdophane—are given (Fig. 3.8).

Splitting of band ν_a (PO_4) 1056–1010 cm^{-1} for ningyoite (and 1060–1009 cm^{-1} for rhabdophane) indicates the presence of non-equivalent PO_4 tetrahedrons in these minerals structures. The main difference of ningyoite from rhabdophane by IR-spectra is the multiformity of water (three types) in ningyoite (diapason νH_2O = 3000-3500 cm^{-1}) and increased content of weakly bound (adsorbed?) water (3380 cm^{-1}). Shoulder at 3530 cm^{-1} in IR ningyoite spectrum apparently refers to OH-ions oscillations. Absence of symmetrical P-O-valence oscillations at 953 cm^{-1}, typical for rhabdophane, indicates absence of strong distortions of PO_4 tetrahedrons in ningyoite structure.

Thus, N.V. Chukanov revealed different forms of OH-groups in ningyoite structure by IR-spectra. Symmetry of PO_4-tetrahedron in ningyoite and its asymmetry in rhabdophane is established. In order to localize light atoms positions in natural ningyoite structure, during further structural analysis, it is necessary to take into account the data presented (summarize) herein, as well as to take into account possible isomorphic substitutions in the idealized ningyoite formula.

3.1.2.3 About Uranium Valence in Ningyoite

T. Muto and co-authors studied enriched U-ore concentrate. Based on the data of its microchemical analysis, IR spectroscopy and radiography, they came to conclusion

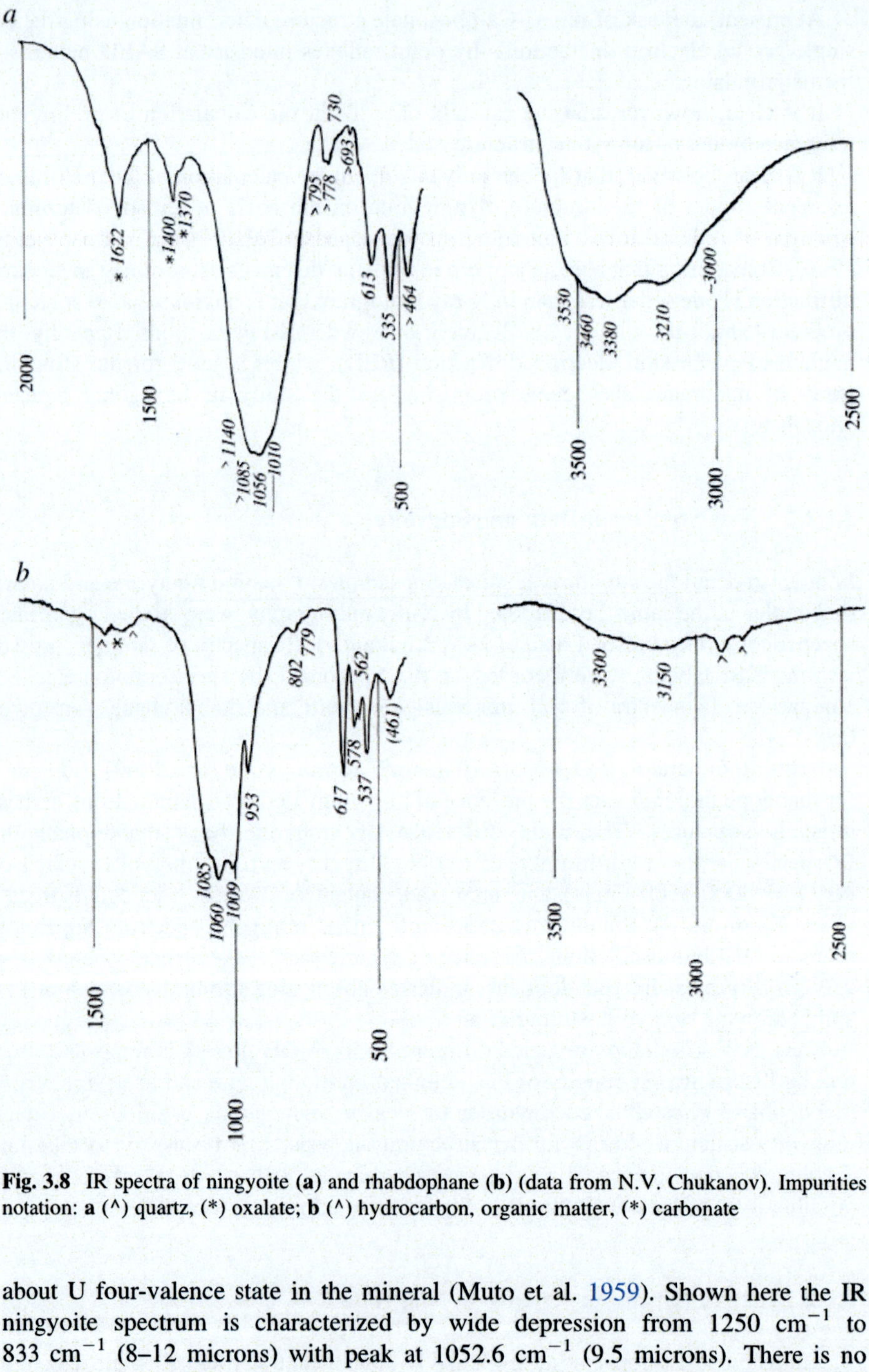

Fig. 3.8 IR spectra of ningyoite (**a**) and rhabdophane (**b**) (data from N.V. Chukanov). Impurities notation: **a** (^) quartz, (*) oxalate; **b** (^) hydrocarbon, organic matter, (*) carbonate

about U four-valence state in the mineral (Muto et al. 1959). Shown here the IR ningyoite spectrum is characterized by wide depression from 1250 cm^{-1} to 833 cm^{-1} (8–12 microns) with peak at 1052.6 cm^{-1} (9.5 microns). There is no presence of uranyl-radical peak in this region, which would be clearly manifested at wavelength 943 cm^{-1} (10.6 microns). It is this fact that allowed Japanese authors to conclude that ningyoite is 4-valence U mineral, because all known hexavalent uranium minerals contain uranyl radical. One of additional arguments in favor of uranium tetravalency Japanese authors considered identity of ningyoite and rhabdophane X-ray-grams; on this basis, the conclusion about their structures similarity was made and formula $CaU(PO_4)_2 \cdot 2H_20$ was proposed.

In mineralogy it was believed for many years that U^{5+} could not create stable compounds, so there are no pentavalent (quinquivalent) uranium minerals in nature. Such an opinion existed before the first publication of Yu.A. Teterin et al., where the presence of U^{5+} (up to 70%!) in natural uranium oxides was reported (Teterin et al. 1981). This publication was occasion to determine uranium valence in our sample of iron-free ningyoite. X-ray photoelectron spectroscopy (XPS) study performed by Y.P. Dikov was carried out on samples of finely worn powder (ningyoite and organics). According to Y.P. Dikov, obtained results indicate predominantly pentavalent U form existence in ningyoite structure (Belova et al. 1985). The treatment of obtained XPS spectra was based on data obtained for uranium oxides in the paper (Teterin et al. 1981). However, the insufficient study of these natural U-oxides samples, taken as reference, leaves doubts in uranium pentavalence in ningyoite.

Several publications are known about the existence of U^{5+}containing minerals, but the question of unambiguity of such uranium valence state determination in ningyoite remains open.

Later, data on synthesis of pentavalent uranium U^{5+} stable compound, and then on finds of natural mineral $CaU^{5+}(UO_2)_2(CO_3)O_4(OH)(H_2O)_7$ were published (Burns and Finch 1999). The proof of uranium pentavalence in this aqueous calcium-uranium uranyl-carbonate is obtained in the mineral structure determining. It is crystallochemical evidence, and it is based on the polyhedron geometry and on sum of valence bonds of one of the U positions in structure, that requires presence of U^{5+} for electron-neutral balance of valences.

In Croatia scientists' publication (Kniewald et al. 2002) for hydrogeochemical and thermodynamic control of organic matter in formation of various U forms in aqueous solutions and U-ores, calculation of thermodynamically possible reactions was carried out. Here, the concentration of organic matter is considered as parameter that defines the system Eh potential (along with radical free energy and oxidation degree) in aqueous solutions. The presence and geochemical significance of transition state of uranium U^{5+} is emphasized, which was taken into account in models predicting redox speciation of uranium and so in reactions of its mass transfer in hydrothermal solutions. Knowledge of uranium redox reactions is considered critical for explaining redox front in uranium-sedimentary mineralization. The authors' hypothesis of U^{5+} stability in the solid phase, in contrast to the aqueous solution, takes into account the equilibrium of this uranium form and thermodynamic stability in slightly oxidized and oxygen-free environment. The address of Croatian geochemists to U^{5+} in the study of the regularities of uranium mineral formation in sedimentary strata is very important. It is reflects both the growing interest in this poorly studied form of uranium and the recognition of the leading role of organic material.

3.1.2.4 Chemical Composition

Taking into account the rare-earth elements (REE) entry into structure, Japanese mineralogists proposed formula $Ca_{1-x}U_{1-x}$ $REE_{2x}[P(O,OH)_4]_2 \cdot 1\text{–}2\ H_2O$ based on results of "wet" chemical analysis of enriched ore fraction (Muto et al. 1959).

According to our data obtained on the EDS analysis of individual microcrystals in ATEM, it was found that in composition of ningyoite microcrystals from reference standard sample from Japan REE is absent. Their fixation in Japanese researchers study is due to the presence other disperse REE minerals in these samples. This was confirmed later by data on the absence of correlation between U and REE in elements distribution patterns on probe (Boyle et al. 1981).

In reference sample composition only characteristic elements of this mineral have been revealed: P, U, Ca (Fig. 3.9). When recalculated to atomic ratios, well analyzed uranium mica—meta-autunite—with the same elements in composition was used as reference (Belova et al. 1978a). From particle to particle, atomic ratios fluctuated in the following ranges P: U: Ca = 2: 0.6–0.9: 1.04–1.1. Statistically, the Ca content is some more compared to U, which is characteristic feature of ningyoite. This is also confirmed by probe analyses results of ningyoites of Bohemia, performed later in scanning analytical EM (Scharmova and Scharm 1994). In the general case, uranium and calcium amount in ningyoite composition is the same, it is characterized by the ratio (U + Ca): P = 1:1, which corresponds to the ratio U: P = 1: 2.

Taking into account some excess of Ca in comparison with U and obtained atomic ratios, the following formula is proposed: $Ca_{2-x}U_x[P(O,OH)_4]_2 \cdot nH_2O$. This formula, also adopted in the publication (Boyle et al. 1981), is convincingly confirmed by numerous EDS testing of ningyoites (Scharmova and Scharm 1994). Based on the assumption of isomorphic calcium replacement by divalent iron in Fe^{2+}-containing ningyoite structure, the following formula is proposed: U_x (Ca, Fe)$_{2-x}$ [P (O,OH)$_4$]$_2$ $\cdot$ n H_2O, where $x \leq 1$ (Belova et al. 1978b).

The studied ningyoites compositions demonstrate the variability of cationic composition (Doinikova et al. 1993). The diagram reflecting the regularities of cationic changing is proposed to detail such transformations, considering that anionic (phosphate) component this mineral's composition must to be constant factor (Fig. 4.5, Chap. 4). The area of compositions, which were most often found among the studied ningyoites, is represented on the diagram by shading.

The chemical composition studies of ningyoite by X-ray spectral microanalysis (RSMA) method «is a complex, not yet fully solved methodological problem, because ningyoite belongs to the category of thermo-unstable hydrous minerals that change their composition ... under the influence of electron beam" (Dymkov et al. 1986). Prolonged exposure of finely focused probe beam leads to water removal and increased concentrations of its elements, and then to the formation of a burnup crater on its surface. In given results of quantitative analysis of ningyoite from hydrothermal veins the sums of analyses are sharply underestimated (usually 61–69.5 wt %), which is typical for mineral with such high water content (up to 20÷23%).

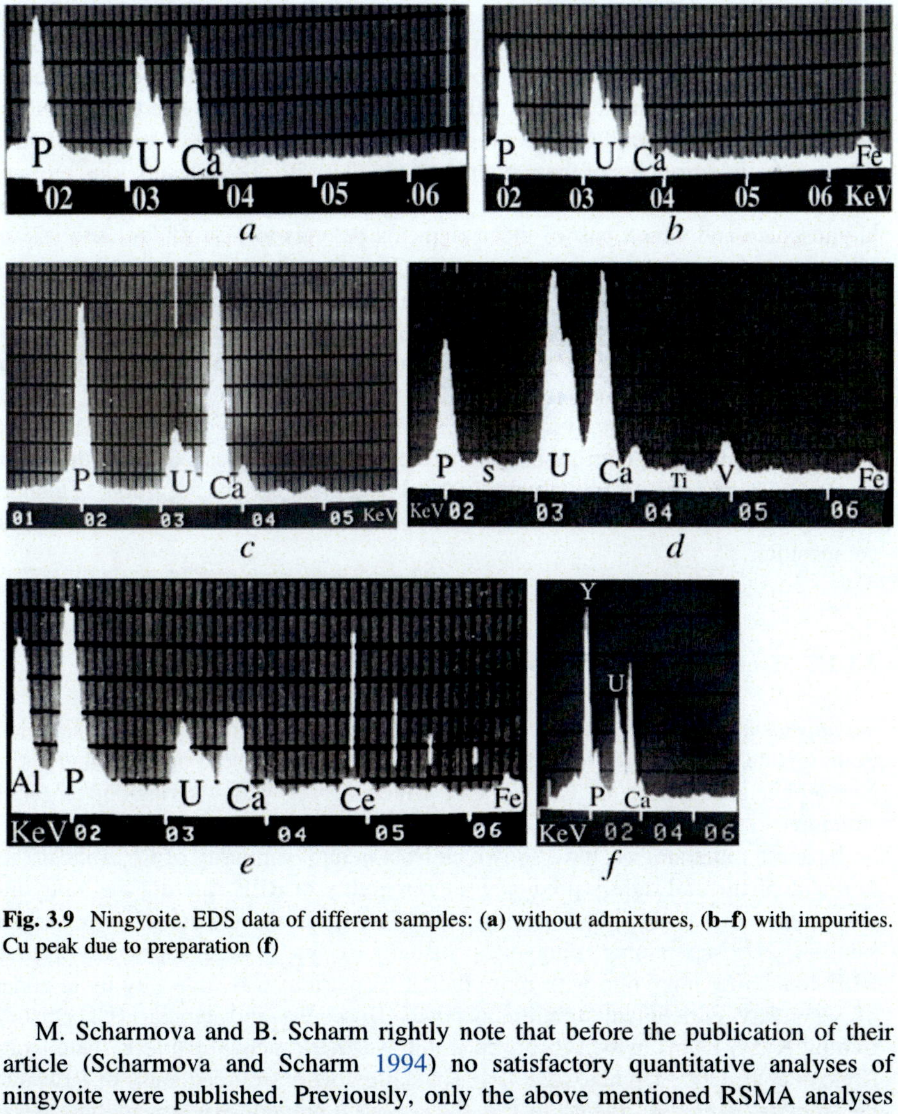

Fig. 3.9 Ningyoite. EDS data of different samples: (**a**) without admixtures, (**b–f**) with impurities. Cu peak due to preparation (**f**)

M. Scharmova and B. Scharm rightly note that before the publication of their article (Scharmova and Scharm 1994) no satisfactory quantitative analyses of ningyoite were published. Previously, only the above mentioned RSMA analyses of hydrothermal ningyoite and result of chemical analysis of enriched ningyoite fraction made by its discoverers were known in the literature. The Czech mineralogists performed about 200 analyses of ningyoite and brockite (Th-analogue of ningyoite, see Chap. 4) on samples from different parts of uranium-ore region of Northern Bohemia (Czech Republic). The analysis was performed in scanning CamScan-4DV with LINK AN 10.000 probe. Th-containing ningyoites with different amounts of thorium in the composition were quite often encountered, which allowed the authors to clearly demonstrate the continuity of the compositional field between the end members of the mineral series of ningyoite-brockite on the diagram Ca-U-Th (Fig. 4.6, Chap. 4). This diagram also demonstrates the "prevalence of Ca over U+Th" previously established in our studies. In the table of 26 analyses, where CaO, UO_2, ThO_2, P_2O_5, Fe, S, Al_2O_3, SiO_2, TiO_2, ZrO_2 were determined, the difference in sums (Σ from 72 to 95%) is related to the rough polishing due to polymineralicity of samples and cavities in mineral aggregates. The Czech authors note that all analyses in the table are presented as mixture and are not recalculated due to close intergrowths of these phosphates with pyrite. The question that remains unresolved: which part of Fe and S represents the isomorphic admixture in U-phosphate and which part of these elements belongs to Fe-sulfides. Analysis of ningyoite-brockites for the presence of CO3-ions in the composition (Microspec WDX-2A) showed significant differences in its content. These results allowed the Czech authors to explain the lack of phosphorus in the composition of ningyoites and, accordingly, the necessary compensation of charges by partial replacement of PO_4-ions with CO_3- and/or SO_4-ions.

In our ningyoites qualitative elemental composition investigations in TEM, when the contribution of neighboring mineral particles in analysis was excluded, the sulfur presence was recorded extremely rarely (peak slightly above background). This fact is interpreted as probability of isomorphism $[PO_4] \leftrightarrow [SO_4]$ in the anionic part of the composition.

3.1.2.5 Isomorphism of Ningyoite Composition

Isomorphism in cationic part of composition. Repeatedly found in different samples, Fe-ningyoite represents variety of Ca-U-phosphate, where, in our opinion, there is an isomorphic substitution of $Ca^{2+} \leftrightarrow Fe^{2+}$, which agrees with its rhabdophane-like structure.

Japanese mineralogists have shown an isostructural similarity of REE-phosphate to rhabdophane and rightly predicted the possibility of REE entering the ningyoite composition. In our studies, among numerous finds of this mineral in various samples, REE-containing ningyoites (usually Y-type) were repeatedly found. REE-containing ningyoite with more than 10% yttrium was found by us in zones of secondary enrichment uranium deposits Koscheka and Jantuar (Uzbekistan, Central Kyzyl-Kum), which localized in the weathering crust of black shale strata (Belova et al. 1985; Doinikova 2003). Thus, previously assumed entry of REE into the mineral structure (Muto et al. 1959) was proved experimentally. These REE-ningyoites were formed in hypergenic conditions. The presence of REE elements, predominantly the Y-subgroup, is characteristic for these conditions and corresponds to the fractionation pattern revealed into hypergenic rhabdophane (Bowles and Morgan 1984).

The presence of REE in analyses of ningyoites from Northern Bohemia uranium ores is considered as random by Czech mineralogists, such rare finds they attribute to close aggregation with other minerals. Only the presence up to 4% of Y_2O_3 is considered by them to be possible isomorphic occurrence of Y in ningyoite composition (Scharmova and Scharm 1994).

Years of experience studying ningyoites from different deposits, indicates the variability of the cationic composition in its natural formations. Conclusions about developed isomorphism of U-Ca-Th-REE in the cationic part of the composition of ningyoites were made in the work (Doinikova et al. 1993). The possible schemes of isomorphic substitutions in cationic composition of minerals with rhabdophane-like structure are considered here by example of ningyoite. The consideration of this U-Ca-phosphate as typical representative of rhabdophane mineral group allowed clarifying the Systematics among isostructural phosphates belonging to this mineral group. More details on cationic isomorphism will be discussed in the next chapter, in the section on mineral Systematics of the rhabdophane group.

Data on *isomorphism in anionic part of composition* were obtained by studying ningyoite with help of ASEM. According to results of natural ningyoite analysis from Bulgaria, about 4% of Si was recorded. Compressed samples of brown-black mononingyoite powder, with amorphous organics (not polishing), which consist of needle ningyoite microcrystals (up to 2 microns) was analysis. Si presence in this phosphate indicates, probably, the possibility of isomorphic substitution of phosphorus by silicon ($P \rightarrow Si$). ATEM study of numerous ningyoite samples in EDS spectra (Fig. 3.9) showed the presence of impurity elements As, V, very rarely S (on spectrometer sensitivity limit) and Ti. According to data (Scharmova and Scharm 1994), isomorphic replacement of PO_4 anions with SO_4 and CO_3 is permitted.

3.1.3 Tristramite Is Analogue of Ningyoite

In 1983, data on discovery of new calcium-uranium phosphate from rhabdophane group—tristramite—were published; it was find in the old mine dumps of Cornwall in UK (Atkin et al. 1983). Taking into account our study of calcium phosphates of tetravalent uranium, which was almost 10 years old, we could not help but notice the similarity of some optical and physical characteristics of tristramite and ningyoite.

Considering our nearly 10-year, by that time, study of calcium tetravalent uranium we could not help noticing the similarity of number optical and physical characteristics of tristramite and ningyoite.

Refraction and habitus of crystals are similar; both minerals are characterized by weak pleochroism, positive elongation and extinction parallel to elongation, by absence of fluorescence. As in ningyoite, the ratio Ca:U in tristramite is close to 2:1. The given tristramite composition is not perfect in relation to the data of microprobe analysis, as well as to only passingly mentioned "partial chemical analysis". The mineral found in dumps of almost century ago, and even with the closest association of uranium pitchblende and tristramite with pyrite and marcasite. These facts allow us to assume significant oxidation (ocher process) of the sample and presents here secondary new formations of uranium and iron. As a result, atypical pale yellow to greenish-yellow color was observed. The impregnation tristramite of tiny particles of goethite, which was noted by D. Atkin, allowed us to consider that he mistakes when attributing the entire iron (established by probe analysis) to tristramite composition. These data do not allow us to agree with the conversion of all uranium to U^{4+} and iron to Fe^{3+}. The set of most intensive lines of tristramite powder-grams corresponds to the same lines of ningyoite (Muto et al. 1959); according to the specified data (Belova et al. 1985) elemental cell parameters of both minerals are similar. The main difference between tristramite and ningyoite was its belonging to hexagonal syngony. Parametric characteristics of ningyoite (and fact that previously adopted rhombic cell actually characterizes orthohexagonal syngony), which were clarified by us in 1985, raise doubts in the individuality of tristramite. In our opinion, the *tristramite* specified as individual mineral is *ningyoite* with a ratio Ca:U $\approx$ 2:1.

Since the tristramite has not been annulled by Commission on New Minerals of International Mineralogical Association (CNM MMA), it is given in summary table of tetravalent uranium phosphates, but is put in quotes (Table 3.1), as not individual mineral from our point of view (Belova et al. 1987).

3.1.4 Ningyoite Occurrences (Deposit Types, Mineral Associations)

The first discovery of ningyoite was made in quaternary conglomerates (Ningyo-Toge formation), lying on the eroded surface of Cretaceous granites, among paleochannel deposits of infiltration deposit Ningyo-Toge, where this Ca-U-phosphate is represented as the main ore component of unoxidized exogenous ores. According to the hydrogenic deposits classification (Kislyakov and Shchetochkin 2000), Ningyo-Toge belongs to *ground infiltration* subclass (type "*paleovalley").*

In the former Soviet Union, some ningyoite finds were made in various deposits mainly associated with *oxidation zones*. The second find in the world was our discovery of ningyoite in the ores of plast-infiltration deposit Sugraly, Kyzyl-Kum ore district, Uzbekistan (Belova et al. 1978a). In addition to ningyoite, uranium pitchblendes (uraninite) are widely represented in the samples. Uranium ore mineralization at this deposit is manifested in sandstone cement in the form of powdery collomorphic formations of uranium black and yordizit mixture. Here, in the ore-bearing sediments there are no carbonaceous organic residues. According to Kislyakov and Shchetochkin (2000), the deposit ores are nasturane-sulfide; from the side of incoming oxygen waters, before the rich ore bodies, areas with uranophane ores are marked. Such uranium mineralization sequence (mineral zonality) demonstrates regularity of redox processes development in hypergenesis zone, that typical for weathering of endogenous (hydrothermal) uranium deposits, which established by L.N. Belova. According to the same authors, over here the role of the reducing agent is obviously played by "superimposed" disulfides formed before the deposition of uranium blacks.

The third find of ningyoite (our second) was made *in weathering crust* of Kosachinoe hydrothermal field in Northern Kazakhstan (Belova et al. 1978b), where Fe-containing ningyoite was first discovered. Here, the crushed-rock debris and clay-carbonate weathering crusts were formed upon clayey and carbonaceous-argillaceous pyrite-containing shales (see Chap. 2). The richest ore sites are confined to unique secondary enrichment zones, where oxidized areas with carrots iron hydroxides and whitewash rims are changed by re-deposited pyrite and fine uranium mineralization. This corresponds to the "partial oxidation" zone in zonality of sandstone-type stratal-infiltration deposits. Secondary ores of Kosachinoye deposit are composed mainly of Fe-containing ningyoite, with subordinate amount of coffinite and poorly crystallized uraninite. Ore formations gravitate towards layers of organic matter. The location conditions and mineralogical situation show that ningyoite was deposited from slightly alkaline, close to neutral solutions.

Our single finds of ningyoite (among uranium oxides) were made the 80s in unique *weathering crust* (in karst sinkholes form) at Zaozernoye deposit, Northern Kazakhstan. According to N.A. Glagolev's data, under karst sinkhole in upper part of ore body there is oxidation zone, under which there is enrichment zone with developed uranium blacks (manly oxides) in association with marcasite and ningyoite.

Ningyoite is especially widely represented in Bulgaria (Belova et al. 1978a, b, IGEM funds), where along with deposits with a complex composition of ores (ningyoite, coffinite and uraninite) there are deposits with mononingyoite ores (mixture with organic matter). In deposits with predominantly ningyoitic mineralization, the amount of coffinite and uraninite is sharply subordinate. Ningyoite in Bulgarian monomineral ores is represented exclusively by Fe-containing variety (Fig. 3.12). In these ore samples spherical electron amorphous (without diffraction) micron sizes uranium formations has been established in close association with ningyoite. By EDS analysis data, only Zn or Al is present in some spherical particles, while in others (Zn + Al). Absence of sulfur in composition suggests that these spherical particles are zincite (ZnO) or alumina (Al_2O_3). This mineral association of Fe-containing Ca-U-phosphate indicates an alkaline formation medium. According to localization conditions, hydrogenic Bulgaria deposits belong to *ground-* and *plast-infiltration* type.

In Poland, several new minerals (Th-ningyoite and ningyoite-like Fe-Th-phosphates) are found *in* Bogatynia *pegmatites* and *in hydrothermal veins* localized in Lusatian granitoid rocks of Lower Silesia (Kucha and Weiczorek 1980).

In Canada, British Columbia, ningyoite is the main ore phase (Boyle et al. 1981) of shallow "near-surface" infiltration deposits "in modern alluvial sediments in places of unloading of uranium-bearing groundwater circulating along fractured granites"; ore deposits are located directly at the daily surface at depth of no more than 30 m (Kislyakov and Shchetochkin 2000). According to their classification, these deposits with certain conditionality degree belong to the *ground*/subsoil—*infiltration* deposits of *paleovalley (paleochannel) type*.

Under the name of tristramite, ningyoite was discovered in the old mine dumps of Great Britain in the developments of quartz-polymetallic, uranium-pitchblende hydrothermal veins of the late nineteenth century. It was found in close association with iron hydroxides in conditions of active action of modern weathering processes (Atkin et al. 1983).

In Russia, Khiagda Ore Field deposits with ningyoite ores are located in Neogene paleovalleys (paleochannels) embedded in the granitoid Paleozoic foundation (Vitim plateau, Northern Transbaikalia, Siberia). By our data, for the first time ningyoite has been installed at Khiagda deposit in Siberia 1985, in richest sample of dispersed ores, where pitchblende and coffinite wasn't established (Ryzhov et al. 1985). In recent years, our study of seven similar Khiagda Ore Field deposits has shown their ores' ningyoite composition (see Chap. 5). This Ore Field is "gully" network of small ore bodies on Vitim plateau, which includes Khiagda deposit. Deposits are localized in the clarified and gray-colored alluvial phases of terrigenous and volcanogenic-sedimentary rocks, enriched with organic residues. Here, as in all other deposits, U-ore ningyoite-rich mineralization is developed in direct contact with oxidized strata. According to classification (Kislyakov and Shchetochkin 2000), this deposits belongs to the *paleovalley type of ground/subsoil infiltration subclass*. There is a clear connection between uranium-bearing paleo streambeds (of "gully" network) and granite-dome structures, which allows us to consider these granites as source of U-ore matter. It should be noted that framboid pyrite is always present in samples of phosphate black ores (Fig. 2.18, Chap. 2). In such precipitates, idiomorphic outlines and unaffected by the change (or dissolution) of the pyrite microcrystals surface suggest that sulphides are relatively "young", formed not earlier than ningyoite.

Numerous ningyoite finds was in stratiform ores of Cenoman platform formations of Bohemian massif, Czech Republic (Scharm and Hofreiter 1978; Scharm et al. 1980, 1994; Scharm 1993; Scharmova et al. 1993). These aggregations classified by the authors as unconformities sandstone type deposits (Scharmova and Scharm 1994). According to hydrogenic deposits classification of Y.M. Kislyakov and N.V. Shchetochkin, uranium deposits in the plate complex of the Czech-Bohemian massif belong to *ground/subsoil infiltration subclass* formations.

Ningyoite *of hydrothermal* genesis is established in ores of five-metal (U-Bi-Co-Ni-Ag) formation of the classic vein deposit of the Czech massif Horní Slavkov (Dymkov et al. 1986); it also wide occur in ore veins of uranium deposits Jáchymov and Rožna, Czech Republic. In hydrothermal veins of Czech massif, ningyoite was found among late sulfides in the proustite-pyrite nests of the uranium-arsenide lens, mainly in the form of granular masses and nodular crust, growing in the form of nasturane or rammelsbergite; an association with apatite, bravoite, monstrositie, nasturane and coffinite was noted. Here, in different samples, the substitution of nasturane (different stages), coffinite and apatite by ningyoite was noted; according to distribution elements patterns (Camebax) it was found that ningyoite was substituted by bravoite, pyrite, and less often by proustite.

Summarizing mineralogy data on ningyoite manifestations in hydrogenic deposits, it is possible to tell that this Ca-U^{4+}-phosphate as significant admixture is present in uranium blacks' composition, or forms individual U-ore (black) formations (see Chap. 2). Association of ningyoite with coffinite and uraninite (as a rule, in its poor-crystallized form—uranium pitchblende) is characteristic. There is quite often close association with pyrite, up to interlayering, which found in samples from Bulgaria and also fixed by Czech mineralogists in samples from Northern Bohemia.

It is necessary to underlined once again the fact of wide spread of ningyoite in nature. That was confirmed by our many years study of U-ores material composition in various deposits of sedimentary cover. This U^{4+}-phosphate was found in ores of different geographically remote regions in different geological locations.

By the 2000s, it became apparent that deposits with ningyoite ores are not uncommon as previously thought, but represent a new phosphate type of uranium mineralization. To date, it has been determined that ningyoite is the main ore component for a number of uranium deposits in Japan (Ningyo-Toge, Tono), Bulgaria (Momino, Haskovo, Navysen), Canada (Blizzard, Tyee), Kazakhstan (Kosachinoye) and on the Vitim plateau in Russia (Khiagda ore field). It is identified as a subordinate component of ores in the deposits of Uzbekistan (Sugraly), Kazakhstan (Zaozernoye), and in some ores of the Czech Republic's infiltration deposits (North Bohemia). Ningyoite's finds are known in Poland and UK. Numerous finds of this phosphate in the composition of U-ores, as well as results of our long-term study of the uranium deposits of Siberia on the Vitim plateau (see Chap. 4) convincingly show that we are talking about a new type of uranium ores—black phosphate ones. That was identified based on the study of exogenous sandstone-type uranium deposits using ATEM methods.

Analyzing of this new type phosphate black ores conditions localization, we can conclude that all known deposits, where ningyoite prevails in ore composition (according to literature and our data) belong to similar type of formation. Series Deposits of Japan (Ningyo-Toge, Tono) and Bulgaria (Momino, Navysen, Maritsa and Haskovo), small "near-surface" Canadian deposits (Blizzard and Tyee, British Columbia), Russian deposits Siberia (Khiagda ore field) and deposits of Czech-Bohemian massif belong to ground/subsoil—infiltration genetic subclass according to infiltration deposits classification (Kislyakov and Shchetochkin 2000). These are mainly paleochannel ("paleovalley" or "paleo-streambed") type deposits. To the same genetic type, with some degree of conditionality, can be attributed commercial ores Kosachinoe deposit in Kazakhstan, localized in the weathering crust. Here we based on the commonality of processes in hypergenesis region.

3.1.5 About Ningyoite Formation Conditions

What was known before. This section summarizes and analyzes literature data on all known ningyoite finds in comparison with our own data on deposits of former Soviet Union and Bulgaria (Belova et al. 1985, 1986; Doinikova et al. 1993). On the basis of deposits geological-geochemical situation analysis, conclusions were made about geochemical conditions of ningyoite formation; namely about neutral (or close to it) slightly alkaline and slightly acidic character of pH solutions accompanying the formation of ningyoite.

All foreign ningyoite researchers came to common conclusion that key role in its formation is played by phosphate ion activity, but views on the phosphorus origin differ. These different points of view are discussed in paper (Boyle et al. 1981). On the one hand, studies by T. Muto and D. Katayama (Muto 1962) assumed that ningyoite was formed by replacing clastic apatite in alluvial deposits, since apatite is present in poor uranium ore, but rarely found in rich ore. This is not confirmed by our EM observations of apatite grains without any signs of dissolution (change) in close proximity with ningyoite in thin sections. On the other hand, K. Kajitani (1970) suggested that phosphate was of organic origin, having been remobilized from Neogene lake-bog sediments and re-deposited to alluvial sediments in the form of ningyoite. This assumption was confirmed by our further uranium ores studies.

In review geochemical article (Kajitani 1970) on ningyoite genesis study analyzed wide range of Japanese mineralogists publications on study, the only at the time, find of ningyoitic uranium ores. In Ningyo-Toge region, ningyoite is the predominant ore mineral in two of the region's five deposits. Analyzing extensive data on elements distribution, K. Kajitani found that distribution of phosphorus and sulfur was crucial for uranium fixation. There is common P/S ratio for the region's deposits. However, ningyoitic ores are characterized by close correlation of uranium with both phosphorus and sulfur (in pyrite form), which is not observed in the region's pitchblende ores. Ningyoitic deposits and ore occurrences (mineral zones) are characterized by abundance of colloidal organic matter; its distribution is correlated with uranium content, here carbonized plant residues are rare. Ores with a predominantly U-oxide (uraninite, pitchblende) composition and in uraninite mineral zones are dominated by carbonized wood; here correlation between U and S (pyrite) is negative. In coffinite zone there is a lot of silicified wood. K. Kadjitani believes that areas with different forms of uranium-ore mineralization differed by the predominance of certain biological forms in their period of sedimentation (fauna, herbaceous or woody vegetation). It is noted that in all studied samples of ningyoitic ores the content of phosphorus is strictly limited by the content of sulfur. Sulfur concentration in sediments is conditioned by biochemical generation of pyrite by sulfate-reducing bacteria in oxygen-free humid medium (Fe and H_2S interaction is noted). K. Kadjitani believes that small organisms decayed, forming a colloidal organic substance with an abundance of humic substance, which served as a substrate and energy source for the bacteria that form sulfides (pyrite et al). It is assumed that pyrite played major role as uranium reducing agent for ningyoite formation. In addition, since pyrite predominates in these deposits in Japan, he believes that it is sulfate ion that was the predominant strong oxygen radical in deposits waters, and connects pyrite formation with biochemical reduction of sulfate. Thus, when analyzing this review paper, the biogenic factor of ore accumulation comes to the fore.

Since the publication states that sulfates are present in rich ningyoite ores, it seems logical to suppose the predominance of autotrophic microorganisms, whose life activity leads to the restoration of sulfate ions ($SO_4^{2-} \rightarrow S^{2-}$).

The paleochannel deposits in Japan have concentric mineral zonality (from center of paleo-streambed): the center is composed of pitchblende and/or coffinite; further to the periphery is followed by zone of ningyoite-rich mineralization, where coffinite is in subordinate quantities and only in silicified wood residues; behind the ningyoite-rich zone uranium is represented by uranyl minerals. The internal (uraninite) zone corresponds to lower Eh values than surrounding external zones, i.e. mineral zonality corresponds to decrease in degree of regenerative capacity from center to periphery. According to Japanese researchers, from ore bodies periphery to their center stands out: (1) weakly reducing zone where uranium will not be fixed, (2) pyrite enriched zone where ningyoite will be deposited, and (3) zone enriched with a carbonaceous substance with coffinite. In last zone coffinite will be formed, since decay of wood increases pH and, consequently, Si concentration in water increases. K. Kadjitani concluded that various options for uranium mineralization were mainly due to predominant reducing agent of environment, i.e. relative amounts of pyrite and carbonized organic matter in sedimentary strata.

For thermodynamic calculations and construction of stable existence fields of studied uranium minerals, K. Kadjitani used data typical of ground waters of Japan. In his work, total activity of Ca^{2+} ion on all diagrams 10^{-2} (as in majority region underground waters ~10^{-3}–10^{-2}). In addition, it is assumed that phosphate ion activity in natural environment with abundance of organic matter should be less than its activity in solution, that equilibrium with $Ca_3(PO_4)_2$, i.e. below known values. Based on uniqueness of ningyoite finding, it was concluded that it is especially acidic environment for its deposition: "Ningyoite should not be deposited from slightly acidic, neutral or alkaline waters". "If ningyoite had been deposited from such waters, it would have existed in many uranium deposits, as phosphorus is quite common in such deposits, but ningyoite is not" (Kajitani 1970). This conclusion is based on assumption that phosphate ion activity necessary for the formation of uranium deposits will be roughly represented by the total amount (ΣPO_4) in equality: $\log(\Sigma PO_4) = -(2 + pH)$. Thermodynamic ningyoite values were obtained from data based on idealized formula $CaU(PO_4)_2 \cdot 2H_2O$. The stability field diagrams are based on these assumptions. Ningyoite and uraninite stability fields overlap if $\log(Ca^{+2}) + 2\log(\Sigma PO_4) < -19.6$ (Fig. 3.10). Environmental characteristics for the case, when in that waters there is progressive oxidation of pyrite, favoring to ningyoite deposition, are derived from diagrams Eh–pH for some iron minerals. The contours of Eh–pH region of ningyoite stability are accepted by T. Kajitani according to the conditions of groundwater during its evolution inside the primary sediments. Fields of relative ningyoite and uraninite stability in situation when $(Ca^{+2}) = 10^{-3}$, $(\Sigma PO_4) = 10^{-7}$, $(\Sigma U) = 10^{-6}$ are shown in Fig. 3.10. Ningyoite field stability above diapason pH = 3–7.8 is replaced by uraninite field in case when $2\log\Sigma(PO_4) + \log(Ca) < -19.6$. Ningyoite is unstable in alkaline reducing media with $pH > 7.8$ due to increased phosphate compositions solubility in alkaline conditions (Muto 1965). Therefore, Eh-pH conditions are important characteristic of this phosphate, as well as the phosphate ion activity.

As a result, K. Kadjitani concluded that the ore-forming solutions have high Eh values; a significant content (abundance) of Ca^{+2} ion, and uranium deposition depends on the number of reducing agents. Uranium was deposited by a reducing medium created by organic material, microorganisms and pyrite. Manifestations of P

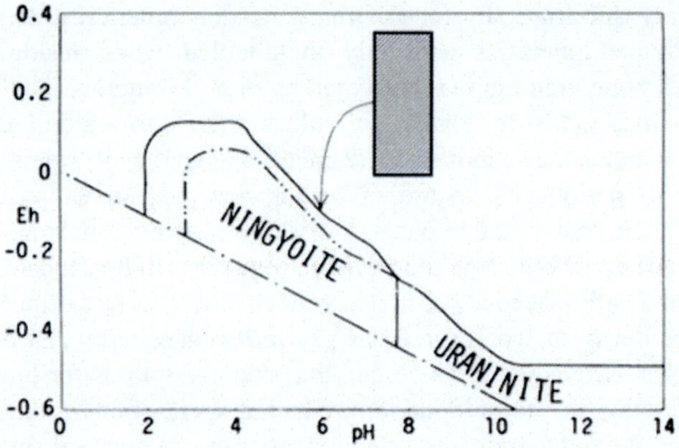

Fig. 3.10 Eh-pH diagram showing the ningyoite and uraninite stability field: $[Ca^{2+}] = 10^{-3}$, $[\Sigma PO_4] = 10^{-7}$, $[\Sigma U] = 10^{-6}$. According Kajitani, 1970; from Boyle et al. (1981). Boundary of uraninite stability field, at $2\log \Sigma (PO_4) + \log Ca^{+2} < -19.6$, indicated by dash-dotted (– ·· –); shaded rectangle—Eh-pH field for most groundwater entering aquifers sedimentary rocks; by arrow – the path of Eh-pH changing of groundwaters that drainage granite massifs (dash-dot – –lower limit of water stability)

and S in the fields in question are directly (or indirectly) related to microbial exposure. Significant sulfate supply in these areas is necessary for the abundant accumulation of sulfur. Erroneous K. Kadjitani's assumption (1970) that pH values required for ningyoite deposition are exceptionally low is refuted by all subsequent discoveries of ningyoites in uranium hydrogenic deposits.

The most interesting are the results of the modern physicochemical conditions study at ningyoitic deposits. Ground waters in Blizzard and Tyee deposits of British Columbia (Canada), which drain the intrusive rocks, underneath the mineralization area, currently contain elevated uranium concentrations (>10 ppb) and abnormally high phosphate concentrations of 80 to 2000 ppb (Boyle et al. 1981). These authors concluded that most of the phosphate is introduced with uranium by groundwater infiltration. It is assumed that ningyoitic deposits are formed *only in the river paleochannels lying on the granite basal complexes*. High phosphate concentrations in alluvial deposits notably exceed quantities that can be explained by simple apatite substitution. In addition, there was no substitution of apatite by ningyoite; we did not observe it. This indicates the secondary role of clastic apatite as source of phosphate in solutions. Our observations fully confirm the validity of these conclusions of the named authors.

According to D. Boyle and his co-authors (Boyle et al. 1981), groundwater draining the granite base is mainly neutral to slightly acidic and has low buffer capacity. It is assumed that optimal conditions for ningyoite formation are achieved where the slightly alkaline groundwater, that leaching basal granite complex and containing high concentrations of uranium and PO_4, filtered through pyrite-containing, rich in organic matter sediments. When analyzing conditions of ore formation in deposits of British Columbia, Canada, it is noted that ningyoite can be formed by marcasite in significantly reduced sediments saturated with humic acids (Tyee deposit) (Fig. 3.2b). It is possible that weakly oxidizing conditions of groundwater, which are favorable for uranyl minerals formation (such as saléeite), locally give way to more reductive conditions. For example, such local changes in conditions are believed to occur as a result of pyrite oxidation to limonite or decomposition of plant material (Blizzard deposit) (Fig. 3.2c). In both Canada and Japan, acid conditions are assumed to be based on pyrite oxidation to citric acid (Blizzard) and the presence of humic acids in ningyoite deposits. According to experimental data (Muto 1965), ningyoite is a low-temperature mineral, which will not be deposited stably at temperatures significantly higher than 25 °C.

It should be especially noted that in all publications where lithology is described, the presence of lacustrine-bog sediments (interlayers) in the ore-bearing rock mass is noted. The analysis of mineralogical data on all known finds of ningyoite leads to the conclusion on the following geochemical conditions of its formation: neutral, or near-neutral, slightly alkaline or slightly acidic character of pH solutions. Our conclusions about the highest Eh values for ningyoite sedimentary deposits, as compared to other types of uranium black mineralization, fully confirm the conclusions of ningyoite discoverers. The active role of atomic hydrogen H^+ in ningyoite crystallization and prevailing weak-gley environment of its formation in hypergenic conditions was noted (see Chap. 4).

The monomineral phosphate ore from the Momino and Navysen deposits in Bulgaria, which samples represented loose dark brown powder (ningyoite and organic matter mixture), clearly demonstrates the significant influence of nature factor on ningyoite ores formation. The contribution of bacterial activity to formation of ningyoite mineralization (secondary ores) can be traced back at Kosachinoe deposit, Kazakhstan (see Chap. 2). Taking into account the close association of this Ca-U-phosphate with organic matter, which is represented practically in all ningyoite finds in nature, we must agree with K. Kadjitani's conclusion about leading role of biogenic factor in ningyoite formation.

All researchers recognize the increased activity of phosphorus in the formation of this mineral. However, the source of phosphorus has remained unclear to date. Recent publications on environmental Mineralogy have helped the author to resolve this issue (see Chap. 6).

What we know now. Our conclusions about geochemical conditions of ningyoite formation were made as on the basis of deposits geological-geochemical situation analysis so on our studies results. Its have led to conclusions about the near-neutral character of pH solutions and the prevailing weak gley situation in the formation of ningyoite (Doynikova et al. 2003; Doinikova 2007). This disproves T. Kadjitani's assumption (1970) of extremely low pH values required for ningyoite deposition from natural waters. However the set of crystallochemical data obtained in this Ca-U^{4+}-phosphate study (analysis of allowable structural isomorphic substitution schemes) indicates the active role of H^+ in its formation (Doinikova 2003) (more detail describe in Chap. 4). A sharp acidity increase can be allowed only in the local (at mineral grain boundaries level) areas in ore-formation zone, which is probably reflected in the observed cases of ningyoite and pyrite overlaying. Such acidity increase may occur as a result of biogenic processes activity.

All publications describing the lithological characteristics of strata with ningyoite ores note the presence of lacustrine-boggy sediments in the host sedimentary strata. Our studies show that phosphate (ningyoite) uranium ores are formed in sandstone hosted basal-channel type deposits mostly in granitoid basement (see Chap. 5).

Genetic Findings from Crystal Chemistry Data Since the main focus of all geological research (including mineralogy and crystallochemistry) is, ultimately, the clarification of ore formation conditions, it was interesting to attempt obtain genetic information through available crystallochemical data. It have been turned out that about *geochemical conditions of* uranous disperse minerals *crystallization* can be judged by their elemental composition, based on the research of A.I. (Perelman 1968). Namely: by peculiarities of elements migration and regularities of uranium concentration in the hypergenesis region. Genetic analysis of crystallochemical data in details is discussed hereafter (Chap. 4). Here, we will only briefly marked the main points and conclusions obtained in the study of ningyoite.

The reasoning was based on the specificity of U^{4+} migration in hypergenic conditions (in aqueous solutions). In this case, the mineral substance dispersity is extremely important. It is this fact allows us to speak about closeness in time of migration processes and crystallization time, and even, with some assumption, of their simultaneity—in the geological sense. In this aspect the available information on elemental composition in mineral components of uranium black ores—coffinites and ningyoites (from various ore samples)—was analyzed from genetic point of view.

Clarify geochemical conditions of ningyoite formation help isomorphic impurities in its composition, which have been established earlier (Doinikova et al. 1993). So, for Fe-containing ningyoite, the simultaneous occurrence into the structure of tetravalent uranium U^{4+} and divalent iron Fe^{2+} (perhaps isomorphic to calcium) indicates a *weakly-gley* reducing transport medium. Other impurities of As, V, Y, S, Ce, Ti captured during crystallization, which were found in different samples (Fig. 3.9), also indicate the predominant *weak gley* environment of its formation in hypergene conditions (Doinikova 2003). It is in this environment that the water mobility/migration of the above elements is allowed simultaneously with uranium U^{4+} and phosphorus by (Perelman 1968).

The analysis of mineral associations in ningyoite ore occurrences allows to judge about the nature of precipitating barriers. Thus, for example, the crystallization of phosphate in a sharply reducing hydrogen sulfide environment (hydrogen sulfide precipitating barrier) is indicated by interbedding ningyoite and pyrite (Figs. 3.11 and 3.12) that was found in the polished samples of the Momino deposit in Bulgaria. The association Fe-containing ningyoite with spherical zinc-containing (oxide?) particles in samples of the Navysen deposit, Bulgaria, where pyrite is absence in ore intervals, allows assuming an alkaline environment of crystallization (Fig. 3.13). The relationship between uranous (black) ores mineralization and geochemical barriers, so as the root causes of extremely small size of crystals (ningyoite, coffinite)

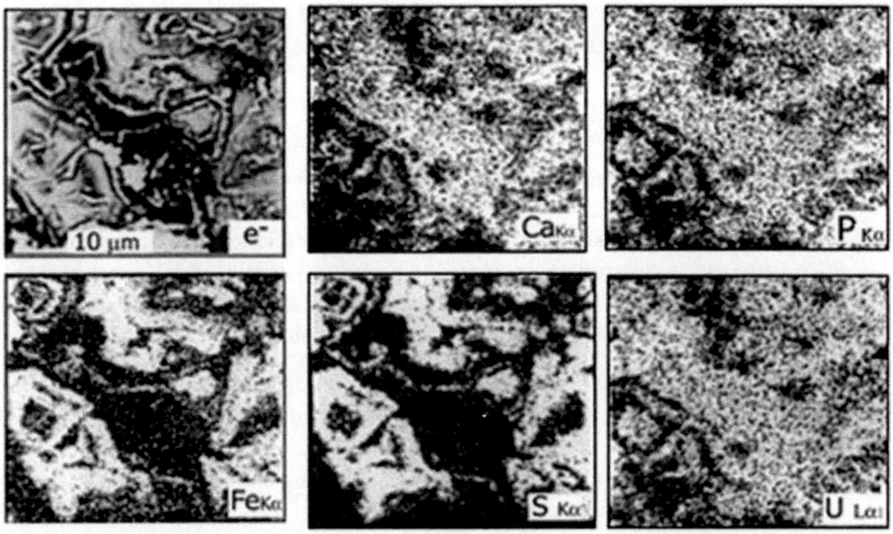

Fig. 3.11 Interlayering of ningyoite-organic mixture (black) with pyrite (light): BSE-image (e-) and Ca, P, Fe, S and U-chemical mapping; polished section (Momino, Bulgaria)

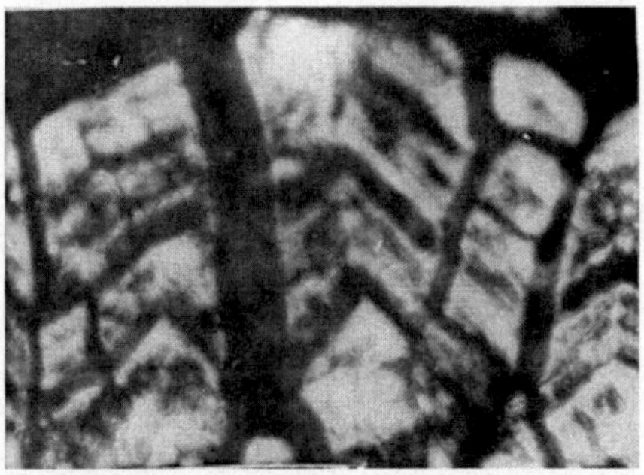

Fig. 3.12 Interlayering in aggregation of ningyoite (black) with pyrite (light); polished section, optic microscope (Momino, Bulgaria)

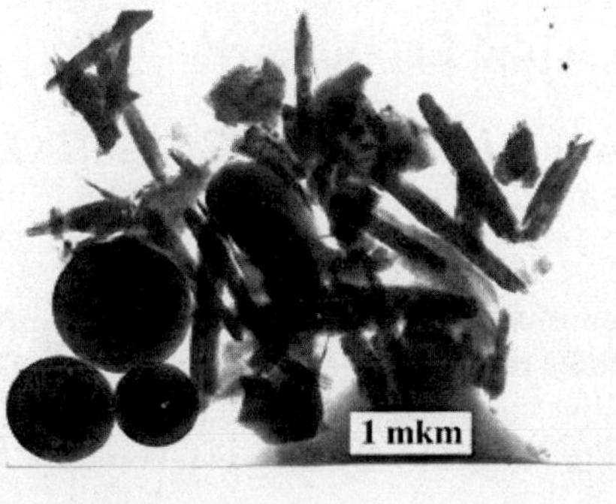

Fig. 3.13 Elongated Fe-containing ningyoite crystals associated with boll-like zincite (?) particles and layered-silicates plates (Navysen, Bulgaria)

and other uranium minerals formations (uraninite) in uranous black ores is discussed further (see Chaps. 4 and 6).

3.1.6 Conclusions

The most widespread mineral from tetravalent uranium phosphates group, ningyoite, which in essence discovered new mineral community—family of U^{4+}-phosphates—was studied. The result of its' detailed crystallochemical study are obtaining number of data that are fundamentally new to uranium mineralogy.

- Previously unknown crystallochemical characteristics of this mineral were obtained: it's belonging to hexagonal symmetry; isomorphism of REE-Ca-U-Th cations in the composition was proved; earlier only assumed REE entry into the mineral structure was experimentally proved; new mineral variety, Fe (or Fe-S)—containing ningyoite, was established.
- Ningyoite has been proved to be widespread in various conditions of hypergenesis region: in weathering crust (Kosachinoe, Northern Kazakhstan), in secondary enrichment zone (Zaozernoye, Northern Kazakhstan), in zones of plast—(Sugraly, Uzbekistan) and ground—(Khiagda, Buryatia) oxidation; and in ore veins of uranium deposits (Horní Slavkov, Jáchymov, Rožna, Czech Republic). This Ca-U-phosphate can be included in U-ores composition to varying degrees, up to monomineral black ores formation.
- Analysis of geological locations and mineral associations for known in the world ningyoite manifestations indicate following its formation conditions: neutral or close to it (slightly alkaline/slightly acidic) solutions; highest Eh values (compared to other types of tetravalent uranium mineralization—oxide, silicate); localization in reductive zone, at redox zonality boundary.
- Obtained by ATEM data served as basis for identifying previously unknown type of uranium ores—ningyoite ones. All known hydrogenic deposits with such ores (in Japan, Bulgaria, Canada, Kazakhstan, Russia) belong to similar genetic type: sandstone type deposits, of plast- or ground (subsoil) infiltration subclass.

3.2 Mineral Group of Lermontovite

3.2.1 Lermontovite: $(U^{4+}_{0.94}\,T\ell_{0.4}\,Ca_{0.02})\,[PO_4]^{3-}$ $(OH)_{1.2}\cdot 0.4\,H_2O$

The discovery of new mineral, Lermontovite, by V. G. Melkov in 1952 laid the foundation for the study of new minerals group—tetravalent uranium phosphates (Melkov and Pukhalsky 1957; Getseva and Savelieva 1957). The discovery, which remains the only one in the world so far, was reported in 1955 at the First Geneva

International Conference on Peaceful Uses of Atomic Energy (Proceedings…, 1958).

Information about lermontovite was limited to brief macroscopic description of the sample, light optics data, and X-ray powder, the reflexes of which were not indicated (i.e., symmetry was not defined). Incompleteness of the chemical analysis for a long time period left the mineral in category of dubious. So at the detailed consideration of some other U^{4+}-phosphates need for additional study of lermontovite. In 1980, sample of lermontovite from the collection of its discoverer was studied. The use of crystal-chemical ATEM methods allowed us to obtain new, more detailed information about this mineral and confirm its individuality (Melkov et al. 1983).

3.2.1.1 Morphological and Optical Characteristics

Morphology. According to the data (Melkov and Pukhalsky 1957; Getseva and Savelieva 1957), the mineral is found mainly in the form of earthy aggregates and veins. On the open cracks walls and other cavities of sulfide vein, the mineral is observed in the form of spheroid aggregates (up to 1.0–1.5 mm in cross-section) of radial-fibrous formation, often covering the crust of Mo-sulphate (mosulite) (Fig. 3.14a–c). At that, Mo-sulfate either does not change at all and retains its black color even if is imbued by lermontovite thinnest cracks system, or is replaced by this U^{4+}-phosphate to form fine-grained aggregate, which consisting of radial-fibrous lermontovite sheafs and the finest dark brown relicts and fragments of mosulite. According to our data (Melkov et al. 1983), the particles of lermontovites in the suspension preparation (lumen image in the electron microscope) have form of plates (Fig. 3.15).

Optical and diagnostic characteristics. It is grayish-green mineral with matte surface and silky shine on the break. The aggregates are fragile. It's density 4.50–4.00 g/cm^3. In UV-rays lermontovite does not luminance; it strongly exposes photoplates in few hours. In thin section lermontovite is transparent, grass-green in color. Radial-fibrous mineral aggregates have concentric-zonal structure. Straight extinction, pleochroism in greenish and greenish-gray tones: $N_p = 1.686 \div 1.690$; $N_m = 1707$; $N_g = 1.724 \div 1.726$; $N_g = c =$ elongation. The mineral from the upper horizons has lower refractive indexes. Lermontovite is easily soluble in nitric and hydrochloric acids; it is not soluble in KON. In sulfuric acid it is dissolved only in Fe^{+3} salts presence (Buryanova 1963; Melkov and Pukhalsky 1957; Getseva and Savelieva 1957).

3.2.1.2 Crystal-Chemical Characteristics

Diffraction data. Lermontovite is represented in suspension preparation by narrow thin plates stretched along the ***c*** axis. The fibrous character of layered structure is clearly visible on EM-transmission pictures (Fig. 3.15a). Point patterns of *electronic*

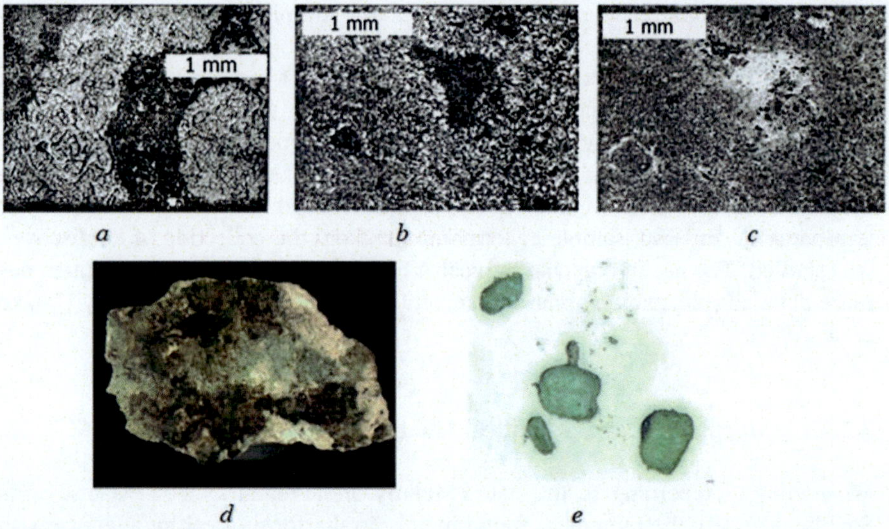

Fig. 3.14 Lermontovite: (**a**) botryoidal aggregates (light) in association with mosulite $MoSO_4$ (dark); (**b**, **c**) fine-grained aggregate of upper named minerals; (**a**, **b**) with analyzer, (**c**) without analyzer; thin section, ×30 (Getseva and Savelieva 1957); (**d**) green lermontovite on a 3 cm matrix (Fersman Museum, Russia; photo Knut Eldjarn/mindat.org); (**e**) grains of green polycrystalline lermontovite (field overview 0.67 mm, photo Pavel M. Kartashov/webmineral.ru). Beshtau U Deposit, Pyatigorsk, Stavropol Krai, Russia. Holotype specimen of V.G. Melkov investigated by L.N. Belova by TEM

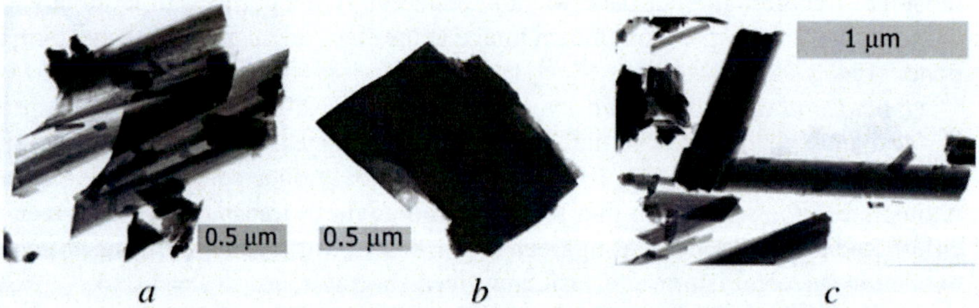

Fig. 3.15 Lermontovite: plate crystal particles (**a**–**c**) in suspension preparat; TEM-image

diffraction from lermontovites plates characterize rather high crystallinity degree; however, under normal device operation the mineral particles are extremely unstable under electron beam. This fact made it impossible to take diffraction pictures from the same particle at different angles. Therefore, almost all electronograms are obtained on different particles of lermontovite at normal or some inclined position of preparation grid to electron beam.

As already noted in methodological Chap. 1, the characteristic elongated mineral particles shape and placing one of crystallographic axes (*c*) along to particles

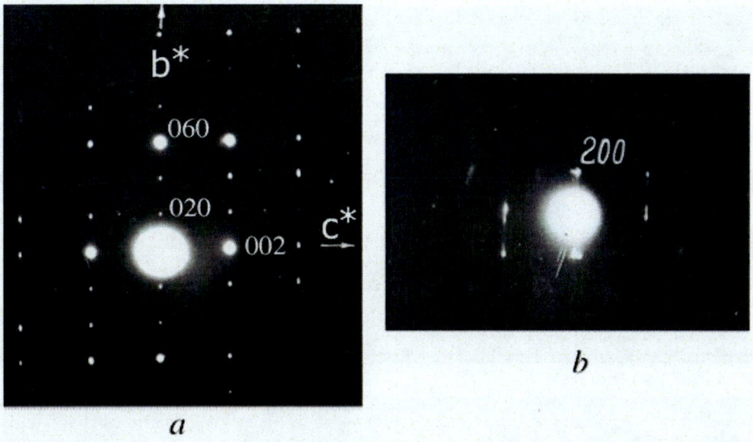

Fig. 3.16 Lermontovite: (**a**) MD-pattern with an orthogonal arrangement of reflexes corresponding to the basis plane of reciprocal lattice (100)*; (**b**) electronogram with basal reflections on a^* axis (first order)

elongation was helped to decoding sets of SAED-patterns. One electron diffraction pattern with orthogonal disposition of reflexes (Fig. 3.16a) was chosen as basis plane (100)*. Others electronograms obtained from different particles were decoding as two series EDPs that have common reflexes lines (00l and 0k0) which were supposed axles ***c*** and ***b***. The rest electronograms can have been divided into groups containing common for each group series of reflexes. Thus, electronograms obtained from different particles, but having common reflexes series *00l* (Fig. 3.17a–d), can be considered as series of different sections obtained by microcrystal rotation around ***c-axis***. Another series of electronograms (Fig. 3.18a–d) was considered as result of rotation particle around the ***b-axis*** or close to it. The third parameter, perpendicular to basal plane, was determined by calculation of set electronograms, which have been "tilted" relative to this basal plane. Some deviation from orthogonality in reflexes spacing on electronograms are considered as result of weak deflection of individual mineral lamellar fibers. Parameters of the mineral have been calculated in assume the elementary cell orthogonality admitting some deviation into monoclinic. Orthorhombic cell parameters determined by point ED-patterns: $a = 8.6 \div 9.8$ Å; $b = 18.6$; $c = 10.1$ Å.

In study sample, except for lermontovite, there are many plate uranium mica (uranite) particles of square forms (Fig. 3.19a). Their identical ED-patterns reflect plane (001)* reciprocal lattice corresponding to perfect mica cleavage plane (Fig. 3.19b). These data made it possible to specify set d-spaces (d/n) lermontovite obtained *by powder X-ray diffraction*. For XRD measurements, sample was prepared by multiple dripping of dense water suspension on glass plate. From the same suspension, only diluted enough, preparations were made for TEM examination. When studying this suspension sample, mica particles are quite often found. It

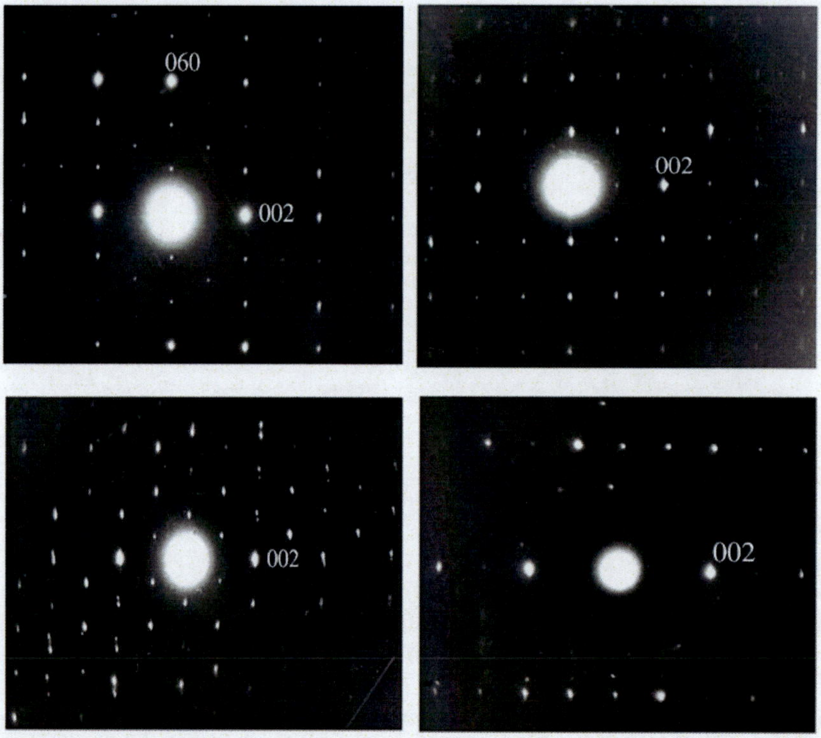

Fig. 3.17 Lermontovite: electron diffraction patterns having shared row ***00l*** reflections, calculated as set of sections upon microcrystal rotation around the c * axis

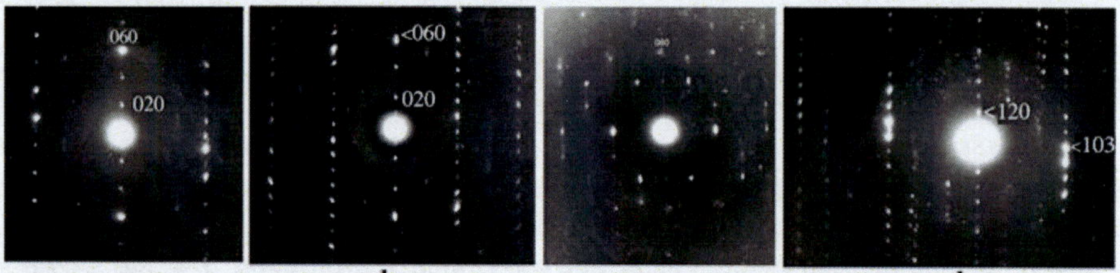

a *b* *c* *d*

Fig. 3.18 Lermontovite: set of electronograms considered as result of rotation of the particle (or plane cutting the reciprocal lattice) around ***b*** * axis (**a**, **b**) or close to it (**c**, **d**)

follows from this that X-ray powder pattern of this sample (Fig. 3.20) is set of peaks lermontovite and uranium mica (uranite).

The most intensive peaks of powder pattern correspond to d-spaces given for lermontovite in Roentgen-metric books of reference (Vasiliev and Kashaeva 1974; Sidorenko 1960). Lermontovite d-spaces set are shown in the table (Table 3.3); it was obtained by mica reflections (shaded in Fig. 3.20a) deduction from total powder

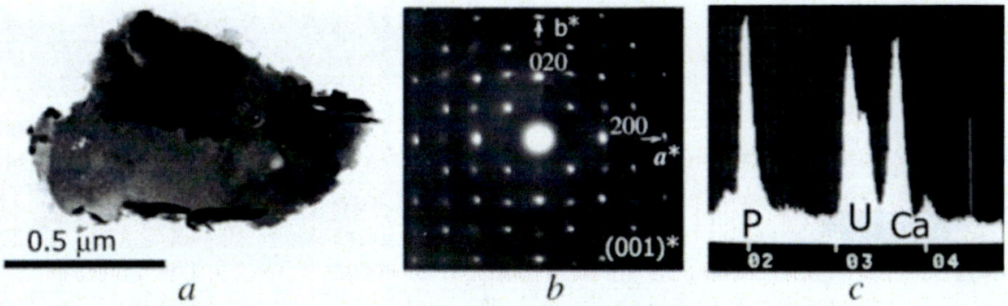

Fig. 3.19 Uranite: (**a**) plate particle from powder lermontovite sample; (**b**) EDS-pattern of this particle corresponding (001)*; (**c**) ED-spectrum of composition

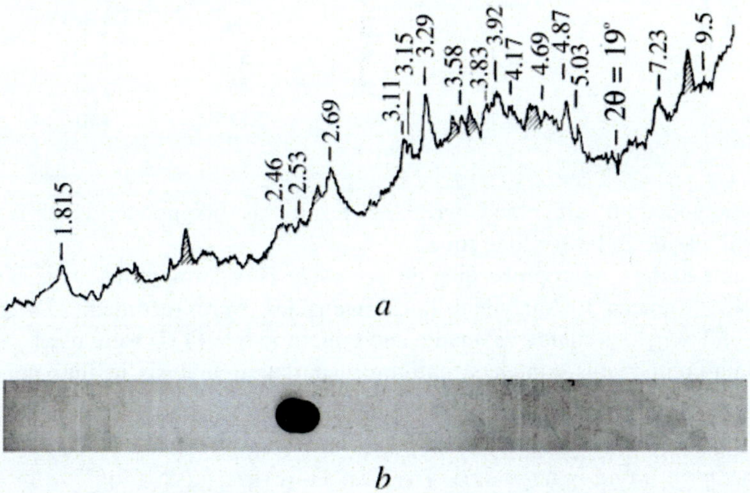

Fig. 3.20 Lermontovite. XRD powder patterns of samples: (**a**) a set of peaks lermontovite and uranium mica (shaded); (**b**) debaegramm

pattern diffraction spectrum. Lermontovite cell parameters values have been specified by powder pattern, and they are: $a = 9.74$; $b = 19.0$; $c = 10.1$ Å ± 0.1 Å. Syngony is rhombic; space group is presumably *Ccca.*

Chemical composition. According to lermontovite discoverers chemical analysis results, this mineral composition is as follows: $T\ell_2O$ 1.55%; CaO 1.00; REE_2O_3 1.67; SiO_2 2.38; UO_2 36.33; UO_3 14.53; P_2O_5 20.40; H_2O 8.72; total 86.58% (analyst A.Ya. Shaskol'skaya). In work (Soboleva and Pudovkina 1957) the mineral formula is presented in following form: $(U,Ca_2TR)_3[PO_4]_46H_2O$. In later work (Getseva and Savelieva 1957) following formula is given: $(U,Ca,TR)_3$ $[PO_4]_4$ $6H_2O$. All uranium in mineral composition is taken as U^{4+} based on green color, which is peculiar to U^{4+} phosphates, and taking into account copper absence in mineral composition (that may be green color source). This chemical analysis have been made on 200 mg quantity of substance weighed for analysis that was polluted

Table 3.3 Lermontovite. X-ray powder diffractometry data

№	*I*	$D_{exp.}$ Å	$D_{calc.}$ Å	*hkl*
1	20	9.5	9.5	020
2	40	7.23	6.92	021
3	50	5.03	5.05	002
4	80	4.87	4.87	200
5	70	4.69	4.70	131
6	80	4.12	4.3	220
7	100	3.92	4.29	041
8	80	3.83	3.93	221
9	80	3.58	3.65	132
10	95	3.29	3.289	222
11	70	3.15	3.167	060
12	70	3.11	3.138	113
13	70	2.69	2.683	062
14	30	2.53	2.525	004
15	30	2.46	2.44	024
16	40	1.815	1.816	174

with molybdenum sulfate and pyrite; its' result was incomplete, which is why the mineral was doubtful for long time.

In our studies, microprobe quantitative analysis of lermontovite was carried out (MS-46 "Cameca"). Analytical lines intensities were recorded: $UM\alpha_1$; $PK\alpha_1$; $CaK\alpha_1$; $T\ell K\alpha_1$; As standards apatite and tallium iodide ($T\ell I$) were used.

Main components content according quantitative analysis in four points (four values are given, respectively) is:

UO_2–66.24; 65.81; 65.48; 65.49%; P_2O_5–19.15; 18.85; 18.34; 17.83%;
$T\ell_2O$—7.65; 11.88; 8.97; 9.34%; CaO—0.24; 0.18; 0.24; 0.42%;
H_2O (by difference)—6.72; 3.78; 6.97; 6.92%.

Mineral composition is accepted as average of four analyses: UO_2–65.63%; TlO—9.46; CaO—0.27; P_2O_5–18.54; H_2O (by difference)—6.10; the sum is 100%.

Discussing issue of lermontovite formula, the following should be noted. With electron microscope help, admixture of uranium mica (uranyl mineral) fixed in the samples. Despite this fact and fact that samples were stored in collection for more than 30 years, uranium oxidation degree that defined in these samples (oxygen coefficient, *o.c.*) does not exceed value *o.c.* $\leq$ 2.2. This suggests that all uranium in the mineral is in U^{4+} form. Therefore, the formula for lermontovite may be presented in form: $(U^{4+}{}_{0.94}\ T\ell_{0.4}\ Ca_{0.02})\cdot[PO_4]^{3-}\cdot(OH)_{1.2}\cdot\ 0.4H_2O$, or idealized—$U^{4+}$ $[PO_4]$ OH· H_2O.

ATEM data on composition. In analysis of individual lermontovite particles it was established that its composition spectrum is represented by peaks phosphorus, thallium and uranium (Fig. 3.21); peak $CaK\alpha_{1,2}$ overlapped by peak $UM_{\alpha}{}^{IV}$. Cu peak in spectrum is due to substrate grid material.

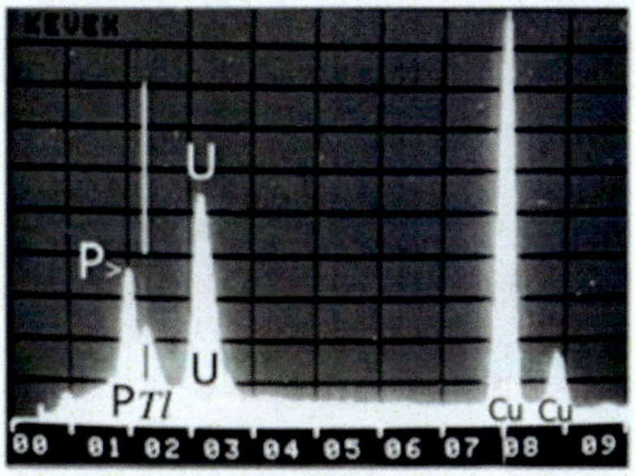

Fig. 3.21 Lermontovite: ED-spectrum of composition. Cu peaks from preparation

Uranium micas (uranites) present in the sample along with lermontovite; its composition mainly phosphate, but there are some particles of mixed phosphate-arsenate composition (insignificant As quantities).

3.2.1.3 Location (Finds, Mineral Associations)

Geological Setting. Lermontovite was found in local secondary enrichment zone at uranium deposit Byk, area Caucasian mineral waters, within one of vein zone sites located in large granite—porphyry boss. The zone contains sulfide vein (~1 m) disturb by tectonic cracks, and less powerful veined bodies up to 10–15 cm. Big veins folded by grey halloysite with small amount pyrite crystals; halloysite also cements brecciated sites sulfide vein.

Literary data (Melkov and Pukhalsky 1957; Soboleva and Pudovkina 1957) show that lermontovite is found only in places located below the groundwater level. Close association of regenerated and residual uranium blacks in places where this uranium phosphate is manifested indicates the presence of secondary enrichment zone here. These facts indicate that Lermontovite is confined to oxygen-free reduction zone of hypergenesis, to the lowest horizons of redox processes manifestation.

The increased acidity of solutions from which lermontovite is formed is clearly indicated by the thallium content of this mineral. According to A.I. Perelman (1968), the Tℓ mobility is typical for highly acidic aqueous solutions. Localization of this mineral directly in sulfide vein (in finest veinlets form) clearly demonstrates the previously established conditions of thallium migration.

Mineral associations. According to data (Melkov and Pukhalsky 1957; Getseva and Savelieva 1957), sulfide vein consists mainly of radial-fibrous marcasite containing arsenic and antimony, and is characterized by constrictions and bulges. Symmetrical-striped structure and large voids in central part of bulges can be clearly traced. The mineral was found in one of sulfide vein bulges in form of thin, up to 1 mm, veinlets and impregnation (Fig. 3.14). Near them on walls of open cracks and of other cavities, nasturane, so-called “residual” uranium black, occur in form of crusts. Besides this aqueous Mo^{4+}+Mo^{6+} sulfate (mosulite), vrbaite $Hg_3Tl_4As_8Sb_2S_{20}$ and lorándite $TlAsS_2$ (As-sulfides of thallium), realgar AsS, opal, evansite $Al_3(PO_4)(OH)_6{\cdot}6H_2O$ (amorphous aqueous Al-phosphate) and, in small quantities, sphalerite and galena occur. Here a few "regenerated" uranium black occur in form of films and thin crusts.

In microcrystalline "earthy" aggregates, lermontovite is located directly in sulfide vein in form of thin, up to 1 mm, gently falling, almost parallel, veinlets. Lermontovite is also found in veins which penetrate into altered granite-porphyry on hanging side.

3.2.2 *Vyacheslavite:* $U^{4+}(PO_4)(OH)\cdot nH_2O$

This new mineral is named after Russian mineralogist Vyacheslav Gavrilovich Melkov, the discoverer of natural phosphates of tetravalent uranium. The mineral was discovered in 1983 in sample from the collection of L.I. Lizorkina, mineralogist Krasnokholmskgeologiya (Belova et al. 1983, 1984a, b). The second discovery of Vyacheslavite in 1990 in North Bohemian ore district of Czechoslovakia was reported in publication of Czech mineralogists Marta Scharmova and Bogdan Scharm with co-authors (Scharmová et al. 1990).

Informative crystallochemical data on the mineral that detected by us in small quantities were obtained only due to use of ATEM complex methods. Vyacheslavite study was carried out with parallel comparative study of tetravalent uranium phosphate Lermontovite.

3.2.2.1 Morphological and Optical Characteristics

Morphology. Vyacheslavite—new phosphate of tetravalent uranium—in the form of dense green aggregates is allocated on quartz crystals in open cavities of late quartz veinlets. Crystalline structure of massive vyacheslavite aggregates, up to 1 mm in size, is revealed only at high magnifications under the electron microscope (Figs. 3.22 and 3.23). Massive formations of this uranium phosphate were studied both on transmission and scanning electron microscope. This mineral was also found in powders form on crystals' surface of quartz, pyrite, and less often fluorite; sometimes it is closely associated with pyrite of octahedral habitus (Fig. 3.22). At high magnifications, it can be seen that dense aggregates are represented by elongated lamellar crystals (up to 8 microns long) or their intergrowths (Fig. 3.23g, e). The thickness of the individual crystals does not exceed 1 μm at width to 5 μm. Less often lamellar crystals of this phosphate forms spherulite-like aggregates with wide side faces (Fig. 3.23c). In the find of Czech mineralogists, thin lamellas of Vyacheslavite up to 10 microns in size were found in intergrowths with columnar ningyoite crystals (Fig. 3.3b-B).

Optical characteristics. Dense mineral aggregates are green to dark green and matt luster. The mineral doesn't luminescence. Under optical microscope one can observe elongated tabular crystals of green color, less often isometric ones. Pleochroism in green tones is weak. Mineral is biaxial negative with parallel extinction/

Fig. 3.22 Vyacheslavite: plate crystals on quartz (**a**–**c**) and on the faces of octahedral pyrite crystals (**d**). BSE-images, SEM

fading and positive elongation. $N_g = 1.731–1.729$, $N_m = 1.729–1.726$, $N_p = 1.700$; 2 V—small.

3.2.2.2 Crystal-Chemical Characteristics

Chemical composition. Almost all uranium in this mineral is tetravalent, since less than 6% of total uranium amount is accounted for by U^{6+} (determination is made by L.S. Shulik, according to methodology given in monograph A.K. Lisitsin 1975). Vyacheslavite EDS data obtained directly in electron microscope, on individual isolated mineral particles is represented by the peaks U and P (Fig. 3.24); CaK_{α} peak is covered by peak UM_{α}^{IV}. Its' chemical composition of was determined for two varieties that differ in color shade: green and dark green. This was performed on x-ray microanalyzer MS-46 Cameca; standards—UO_2 and apatite; analytical lines UMα, PKα and SKα. True concentrations calculation was carried out by ZAF method under Puma program (Boronikhin and Tsepin 1980).

Fig. 3.23 Vyacheslavite: spherolite-like aggregates of crystals (**a**–**c**) and intergrowths (**d**, **e**). BSE-images, SEM

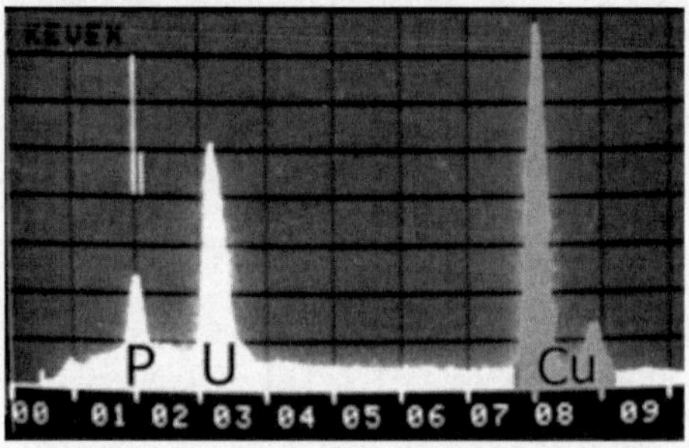

Fig. 3.24 Vyacheslavite's ED spectrum. Peak of CaK_{α} is overlaps by peak $UM_{\alpha IV}$. Cu peaks from preparation

Composition of *green* Vyacheslavite variety (2 analyses data, wt %): UO_2 69.75, 67.63; CaO 0.5, 0.55; P_2O_5 16.9, 17.1; H_2O (by difference) 12.85, 14.72. Empirical mineral formulas, respectively: $(U_{1.08}Ca_{0.04})_{1.12}\cdot(PO_4)\cdot(OH)_{1.4}\cdot 2.3H_2O$ and $(U_{1.06}Ca_{0.04})_{1.1}\cdot(PO_4)\ (OH)_{1.3}\cdot 2.7H_2O$. Composition of *dark green* Vyacheslavite (average of 6 analyzes, wt %): UO_2 76.98, CaO 0.55, P_2O_5 17.07, H_2O (by difference) 5.4%. Empirical formula $(U_{1.18}\ Ca_{0.04})_{1.22}\ (PO_4)\cdot(OH)_{1.82}\cdot 0.35H_2O$ corresponds to analysis results. Significant uranium excess compared to ideal formula has not yet been explained. Dark green, sometimes almost to black, color of this Vyacheslavite variety allows us to assume the presence of non-fixed UO_2 form in it. Significantly smaller water amounts were taken into account in the ideal formula derivation, where $n = 0$–3.

Fig. 3.25 Vyacheslavite. Plate crystal particles (TEM-images in suspension preparations)

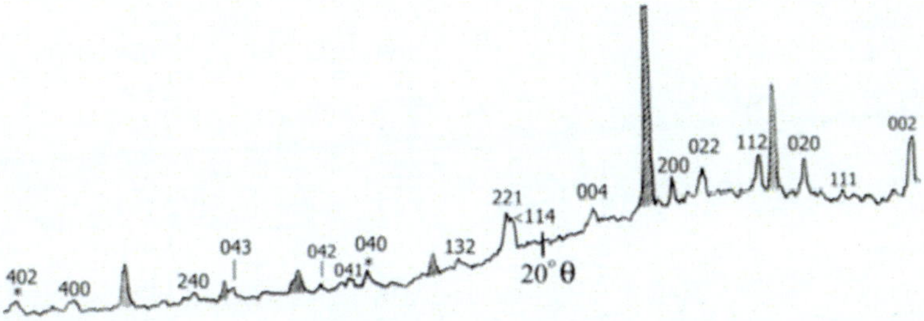

Fig. 3.26 Vyacheslavite. X-ray diffractogram (quartz peaks—shaded; * phosphate peaks coinciding with quartz peaks)

When formula calculating, preference is given to lighter shade mineral variant, which has ratio U: P close to one. Generalized Vyacheslavite formula: $(U_{1.06}Ca_{0.04})_{1.1}(PO_4)(OH)_{1.3}\cdot 2.7H_2O$.

We proposed the averaged formula of Vyacheslavite: $(U^{4+},Ca)\cdot(PO_4)OH\cdot nH_2O$, where $n = 2.5$. Vyacheslavite is an almost pure uranium phosphate with very minor calcium amount. This new mineral density determined by micro method (mineralogical laboratory of IGEM) is 5 g/cm^3; estimated density for idealized formula with $n = 2.5$ is 5.02 g/cm^3.

Diffraction data. Series of SAED patterns representing different cross-sections of reciprocal lattice, including the coordinate ones, were obtained at individual microcrystals when they rotated around crystallographic axes, which made it possible to determine the rhombic syngony and space group of Vyacheslavite (Chap. 1, Figs. 1.5 and 1.6). The use of goniometer was decisive importance for diffraction SAED investigation of new U^{4+}-phosphate and, ultimately, for proof of its individuality.

In suspension preparations this mineral was observed mainly in isometric form, less often plates particles with slight lengthening along ***c*** axis (Fig. 3.25). SAED microdiffraction patterns (MDP) detected during normal incidence of electron beam on preparation most often displayed plane a^*c^* of reciprocal lattice. This suggests that new phosphate has cleavage or predominant plates of (010) (Fig. 1.5). Parameter ***b*** value is determined by number of basal reflections *0 k0* (Fig. 1.6) and specified by XRD (Fig. 3.26). According to SAED data (Chap. 1), the space group of Vyacheslavite is one of following: *Cmcm, Cmc2₁* or *C2cm.*

It was not possible to make pure monomineral selection for study when obtaining powder X-ray because of small mineral formations size. In hand picking sample

Table 3.4 Vyacheslavite: powder x-ray calculation results. Dron-1.5; Fe Kα

I	$D_{meas.}$ Å	$D_{calc.}$ Å	*hkl*
10	6.19	6.19	002
2	5.03	5.05	111
6	4.56	4.56	020
6	4.13	4.12	112
5	3.68	3.67	022
2	3.48	3.48	200
3	3.04	3.045	004
5	2.71	2.71	114
7	2.69	2.699	221
2	2.53	2.54	132
3	2.28[a]	2.28	040
1.5	2.24	2.24	041
1	2.17	2.199	042
2	1.993	1.996	043
2	1.92	1.91	240
2	1.74	1.74	400
2	1.67[a]	1.67	402

[a]Overlapping reflections of vyacheslavite and quartz

there was quartz, on which surface the growing green phosphate was well distinguished by its color. Quartz peaks on diffractogram are shaded; phosphate peaks coinciding with quartz peaks are marked with asterisk (Fig. 3.26). New phosphate elementary cell parameters with regard to XRD data are follows: $\boldsymbol{a} = 6.96 \pm 0.01$ Å, $\boldsymbol{b} = 9.10 \pm 0.01$ Å, $\boldsymbol{c} = 12.38\ 0 \pm 0.01$ Å, Z = 8. Based on these parameters, all XRD peaks are indicated (Table 3.4).

Recent new data on the vyacheslavite structure is provided in the work (Steciuk et al. 2019). The crystal structure of this U(IV)-phosphate has been solved from precession electron diffraction tomography (PEDT) data on the natural nano-crystal. It was established: vyacheslavite is orthorhombic; space group *Cmca*; a ≈ 6.96 Å; b ≈ 9.07 Å; c ≈ 12.27 Å; V ≈ 775 Å; Z = 8. According to these authors, its structure is a complex heteropolyhedral framework consisting of sheets of $UO_7(OH)$ and PO_4 polyhedra, that running parallel to (001) and interconnected by additional PO_4 polyhedra. There is an (OH) group associated with the U(IV) polyhedron. The question of H_2O presence within this mineral framework has been addressed to DFT calculations (structure refining using density functional theory—DFT). Results of such calculations allowed authors to believe that vyacheslavite does not contain any significant amount of molecular water (at room temperature).

3.2.2.3 Location (Finds, Mineral Associations)

Vyacheslavite was found in secondary enrichment zone of Koscheka deposit of complex genesis in Uzbekistan (Central Kyzyl-Kum). There are associates with quartz, pyrite, less often fluorite. The second find of Vyacheslavite was made in ore district of Northern Bohemia, Czech Republic, in stratiform ores of Cenomanian platform formations of Bohemian massif (Scharmova and Scharm 1994). Here it is found in the form of close intergrowths with mineral of ningyoite-brockite composition (Fig. 3.3b). There's association of Vyacheslavite with pyrite and arsenopyrite. Mineral associations indirectly indicate its formation conditions.

The application for discovery was considered by Commission on New Minerals and Mineral Names of All-Soviet Union Mineralogical Society on February 10, 1982. The Commission on New Minerals and Mineral Names of International Mineralogical Association (CNM IMMA) approved our discovery of new mineral—Vyacheslavite—on May 27, 1983. The sample is kept in Fersman Mineralogical Museum, Russian Academy of Sciences.

3.2.3 *Urphoite:* $U_6(PO_4)_7(OH)_3 \cdot nH_2O$

In the mid-1990s, another new phosphate of tetravalent uranium, mineral with idealized formula $UPO_4OH \cdot nH_2O$, was established according to our crystallochemical studies (Belova et al. 1996) in oxidation zones of Koscheka and Dzhentuar uranium deposits in Uzbekistan (Central Kyzyl-Kum), and the Kanzhugan deposit in Kazakhstan (Chu-Sarysu depression).

Characteristic difference of this mineral from above U^{4+}-phosphates is micaceous, layered morphology of flake crystals, maximum size of which rarely reaches several micrometers. This phosphate was first recorded in early 1980s as component of weak sandy-clay-phosphate ore agglomerate. However, its formations character did not allow for detailed mineralogical studies. ATEM methods has allowed to establish initially only the fact of sharp difference of its structural parameters from those for earlier known natural phosphates U^{4+} described above. The fine dispersity of studied crystalline material, in all finds, as well as insignificant size of formations made it extremely difficult to study it by traditional mineralogical methods. The most informative in the study of samples of this new phosphate were methods of EM.

3.2.3.1 Morphological and Optical Characteristics

Morphology of formation. In Kanzhugan deposit, this mineral is manifested as the main component of weak sandy-clay-phosphate agglomerate of greenish-grey, earthy color. In Dzhentuar deposit, new phosphate is found in form of very fine formations on uranium 'mica' (uranite) surface (Fig. 3.27), especially on Fe-saléeite, less frequently on coconinoite (Al-uranyl-phosphate). Such new formations sharply are prominent by emerald-green shade of dark green, different from known color of mica. That new mineral sharply distinguished by emerald-green shade of dark green, different from known color of uranyl minerals (uranites or uranium micas).

a *b*

Fig. 3.27 Urphoite. (**a**) Plate microcrystals of phosphate; rosette-like aggregates on U-micas surface (Djantuar); (**b**) scaly aggregate of crystals. BSE-image

The picture of relationship between this new phosphate and uranium mica is very similar to that of ningyoite (on saléeite), marked by D. Boyle at Blizzard deposit, British Columbia, Canada, where ningyoite crystals are formed by small cleavage cracks in saléeite (Boyle et al. 1981) (Fig. 3.2c). In Koscheka deposit the mineral is represented by dense small nesting clusters (several mm) or layers (< 1 mm) of dark green, emerald green color, sometimes with pyrite crystals up to 1 mm. On the weathered light surface, composed mainly of halloysite, green formations of new phosphate are represented by microcrystalline spherulites of $n \cdot 10$ μm size (Fig. 3.28).

Morphology (According to EM Data). Lamellar microcrystals new mineral growing on the surface of micaceous uranites (Dzhentuar) form aggregates of up to 10 microns, resembling sockets (Fig. 3.28). According to microcrystals appearance and perfect cleavage presence, it is assumed that its crystal structure is layered. In ASEM study of loose samples from Koscheka deposit, it was found that this phosphate forms spherulites of 20–100 μm (Fig. 3.28a, b) and interwoven-foliated aggregates, and covering surface by interwoven lamellar particles, in close association with layered silicates, as it is in Kanzhugan samples (Fig. 3.28c, d). Dimensions of individual plates and small lamellar particles are several microns (Fig. 3.29).

Optical Characteristics. Mineral optical constants determination is complicated by microcrystals dispersity and disorientation of smallest scales that often stickled together. Refractive indexes are installed not more exactly than $N_p' = 1.707–1.708$, $N_m' \approx N_g' = 1.734$ (Table 3.1). Shine is glass; fracture is shell-like. It should be noted that the refractive indices are quite close for all studied phosphates of tetravalent uranium.

3.2.3.2 Crystal-Chemical Characteristics

Chemical composition. New U^{4+}-phosphate formations of small size are found among predominant number uranyl minerals of oxidation zone in sharply

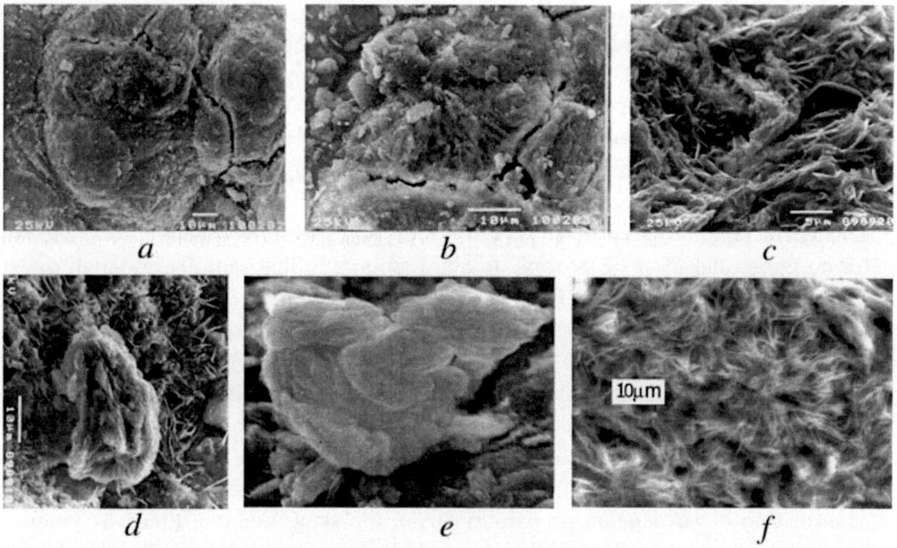

Fig. 3.28 Urphoite. (**a**, **b**) microspherolites (Koscheka); (**c**, **d**) entangled-scaly aggregates (Kanzhugan); (**e**, **f**) rosette-like formation and tangled-scaly aggregates on the surface of loose sample (Koscheka). BSE-image

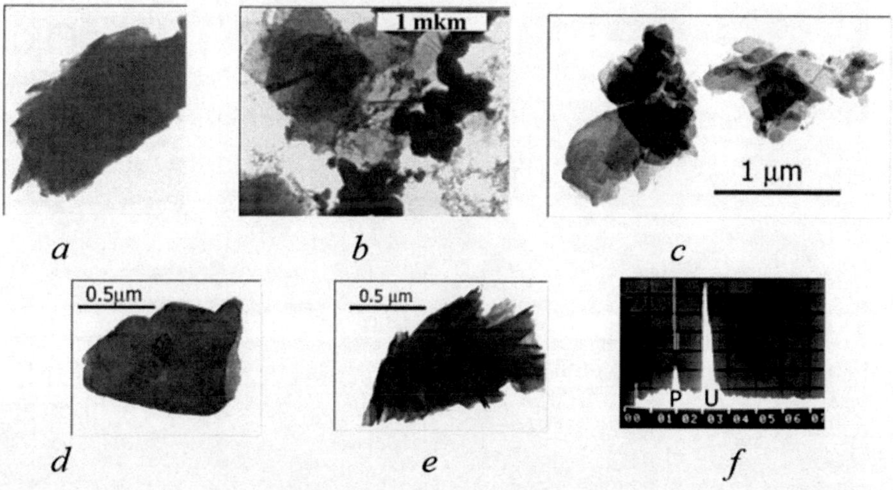

Fig. 3.29 Urphoite. Tabular-scaly phosphate particles in suspension preparation, TEM-images (**a**–**e**); (**b**) in association with nasturane (black rounded particles); (**e**) ED-spectrum of composition

subordinate amount. Yet initially, L.N. Belova assumed tetravalent U form of mineral by its specific shade of green, which is not typical for green-colored uranyl minerals. Due to impossibility to pick out amount of monomineral fraction sufficient for complete chemical analysis, uranium valence in new phosphate was determined by value of oxygen coefficient (o.c.), as this accepted in practice mineralogical work of IGEM. Uranium oxidation degree was determined by domestic national method given in Lisitsin (1975). This method essence is determining hexavalent uranium U^{6+} percentage from total uranium amount in mineral. Value o.c. = 2.0 implies the lowest uranium oxidation degree U^{4+} (absence of U^{6+}) and oxide formula $UO_{2.0}$.

For Kanzhugan ore samples, composed mainly of phosphate material, *o.c.* = 2.15; for highly enriched fraction from Dzhentuar o.c = 2.17 (analyst L.S. Shulik). This corresponds to records $UO_{2.15}$ and $UO_{2.17}$ respectively. That is, U^{6+} accounts for no more than 15–17% of total uranium amount in that samples. A small oxygen factor excess over value 2 for total uranium is most likely due to insignificant admixture of U-V-mica (uranite–tyuyamunite), which is established in these same samples on ATEM data.

ASEM data: elements distribution patterns in these phosphate spherulite-like formations fix only P, U and oxygen (Figs. 3.30 and 3.32b). According to ATEM (SAED and EDS analysis), sample from Koscheka contains insignificant admixture of vanadium uranite tyuyamunite. Admixture of tyuyamunite was established during examination of suspension preparation from the same material that was given for microprobe analysis. ATEM data: insignificant amounts of Fe and V were detected by EDS in some phosphate particles (Figs. 3.31c and 3.32). This sample composition, according to microprobe data (2 probes): CaO 0.76–0.74; P_2O_5 20.62–18.70;

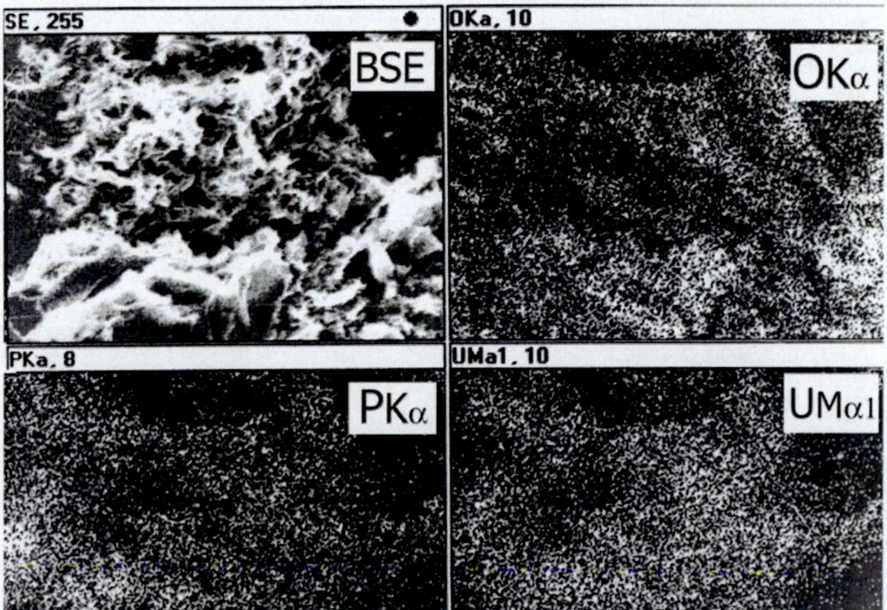

Fig. 3.30 Urphoite. BSE-image and images in characteristic radiation of elements $O_{K\alpha}$, $P_{K\alpha}$ and $U_{M\alpha 1}$

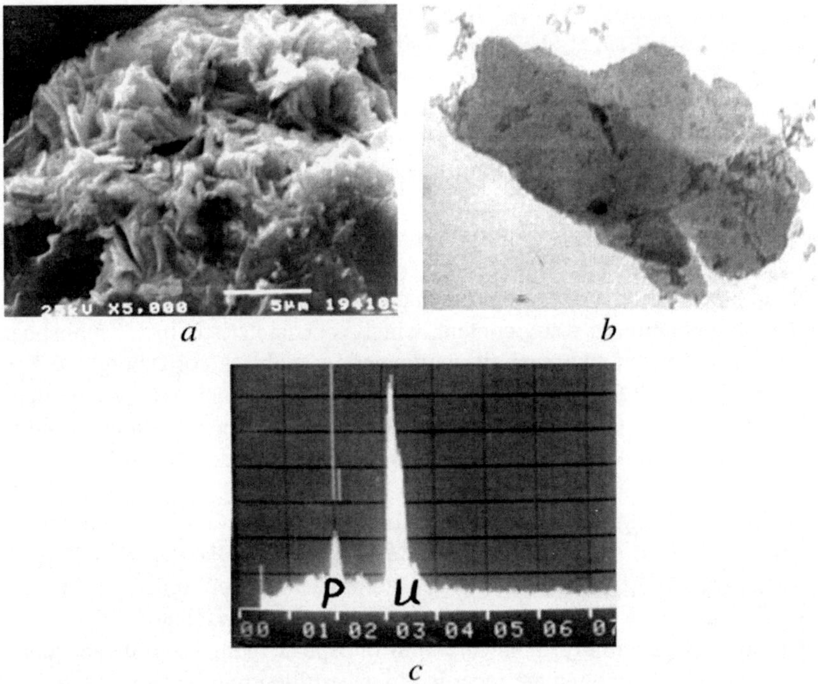

Fig. 3.31 Urphoite. SEM-image (**a**); plate particle in suspension preparation, TEM-image (**b**); ED spectrum composition (**c**); Cu peaks from preparation

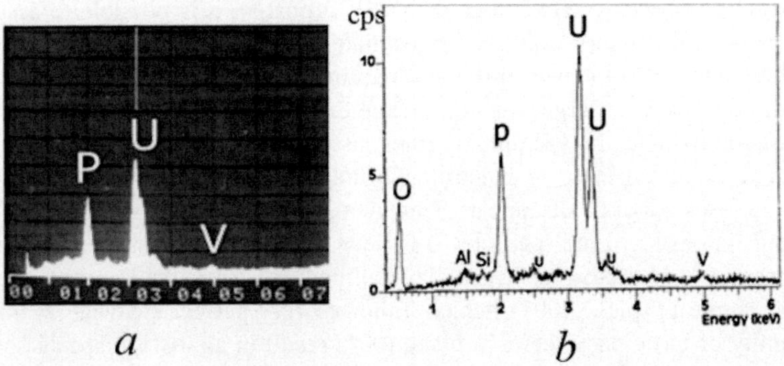

Fig. 3.32 Urphoite. ED spectra of individual phosphate particles in TEM (**a**) and from microcrystals in SEM (**b**)

V_2O_5 2.20–2.27; UO_2 71.82–68.46; H_2O (difference) 4.60–9.83%. Average of 2 analyses: CaO 0.75; P_2O_5–19.66; V_2O_5 2.235; UO_2 70.14; H_2O 7.215%. Standards are UO2 and apatite, analytical lines UM_α, PK_α, CaK_α. Calculation was by ZAF method, PUMA program; analyst V.A. Boronikhin.

Tyuyamunite composition calculated for entire V_2O_5 was subtracted from probe results; subsequent phosphates formulas calculations were carried out for next compositions: UO_2 65.29–61.72; CaO 0.08–0.04; P_2O_5 20.62–18.70; H_2O 2.86–8.03%. Corresponding formulas calculated per one group (PO_4): $U_{0.84}Ca_{0.005}(PO_4)\cdot(OH)_{0.48}\cdot0.25H_2O$ and $U_{0.87}Ca_{0.003}(PO_4)\cdot(OH)_{0.485}\cdot1.6H_2O$. Composition fluctuations in these two analyses are actually related to fluctuation in water content, which is characteristic of U^{4+} phosphates.

Average value of two urphoite composition analyses corresponds to formula: $U_{0.85}Ca_{0.004}(PO_4)(OH)_{0.48}\cdot nH_2O$, where n ≈ 0.3. With integer coefficients—$U_6(PO_4)_7(OH)_3{\cdot}nH_2O$, where n ≈ 2–4. When writing composition formula, it is necessary add hydroxyl group (OH) to compensate for excess positive charge and preserve mineral electro neutrality.

When calculating per uranium unit, the formula shows small predominance of phosphorus over uranium: $U_{0.995}Ca_{0.005}(PO_4)_{1.19}(OH)_{0.42}$ ~ $2H_2O$ and $U_{0.997}Ca_{0.003}(PO_4)_{1.15}(OH)_{0.55}$ ~ $7H_2O$. Preference is given to first variant of calculation, by analogy with compositions of previous U^{4+} phosphates.

Urphoite is essentially mono uranium phosphate with insignificant amount of Ca. Only P, U and oxygen are recorded in spherulite urphoite formations composition on element distribution cards (ASEM) and EDS spectra (ATEM). Its composition is close to ideal formula $UPO_4OH{\cdot}nH_2O$, like Vyacheslavite. This studied new phosphate is also characterized by ratio U:P = 1:1, but with a small phosphorus predominance, while in number vyacheslavite samples deviation towards uranium increase was observed (Belova et al. 1983). Probably, it is possible to assume that this new tetravalent uranium phosphate is polymorphic modification of vyacheslavite, but to clarify this issue requires additional research and, above all, new discoveries with larger mineral formations.

Diffraction data. In vacuum of transmission electron microscope the mineral appeared to be unstable to electronic beam influence: at usual shooting modes weak reflexes quickly disappear. That demanded careful work in mode of low lighting intensity. Fine particles of new phosphate in suspension electron-microscopic preparation have lamellar outlines with smoothed angles; they lie on film substrate by plane (001), which indicates very perfect cleavage at this plane. Proximity of basic parameters in plane (001) result in almost square disposition of microdiffraction reflexes.

These particles resemble uranium mica (autunite) by morphology, elements set and diffraction figures of plane (001)* from cleavage plates (Fig. 3.33, two upper MDP). However, series of electronograms obtained by urphoite particle rotation around coordinate axes a* and b* (Figs. 3.33 and 3.34) allowed to calculate the phosphate elementary cell parameters that different from autunite. These MDP series allowed to recreate planes of reciprocal lattice (100)* and (010)* with help of graphical constructions and to determine location of lattice nodes in graphical

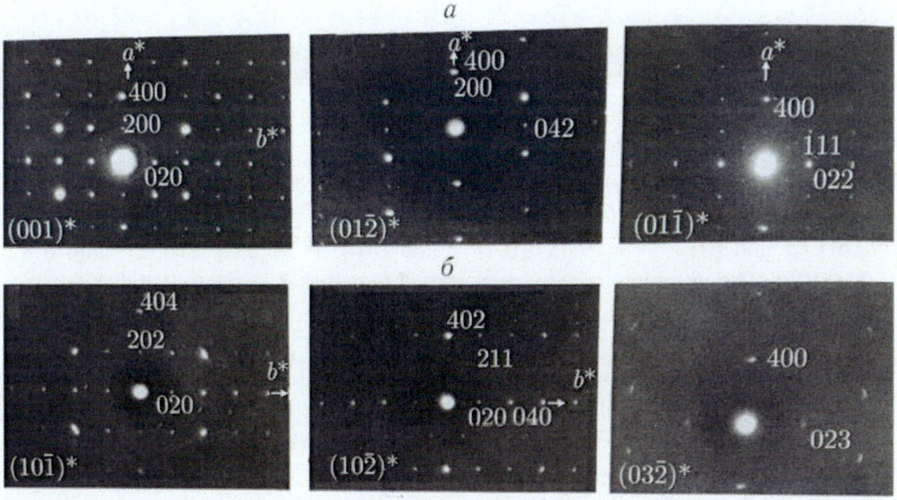

Fig. 3.33 Urphoite. The series MD-patterns obtained by rotating the particle around the coordinate axes $\boldsymbol{a}$* (left column) and $\boldsymbol{b}$* (right column)

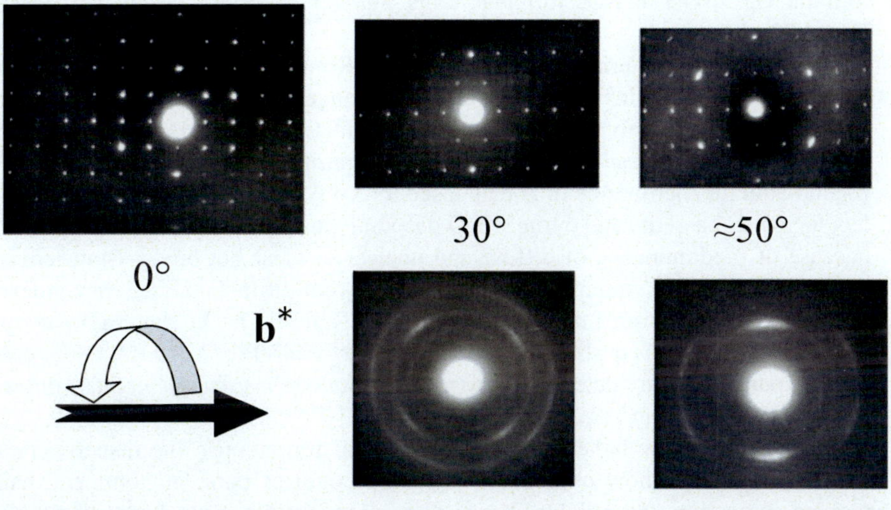

Fig. 3.34 Urphoite. Electronograms obtained by rotating the particle around the coordinate axes $\boldsymbol{b}$* and ring MDPs demonstrating texturing

“reciprocal space”. In contrast to previously mentioned monoclinal syngony (Belova et al. 1996), our subsequent diffraction studies have led to conclusion about orthogonality of crystallographic axes.

The graphically modeled plane (100)* shows rectangular reflexes grid (figure) and symmetrical location of planes (0kl $^{+}$) and (0kl $^{-}$) with respect to plane (001)*

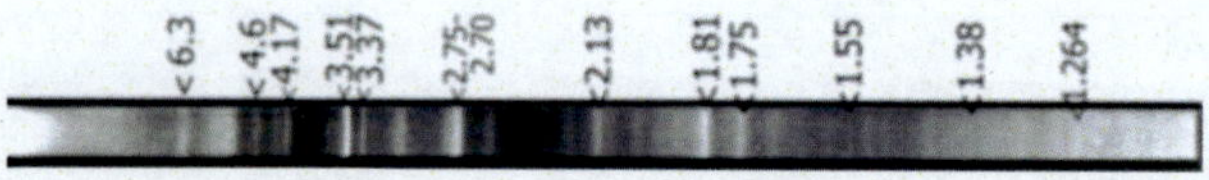

Fig. 3.35 Urphoite (Dzhentuar). X-ray powder pattern (Guinier chamber, $Cu_{K\alpha}$)

and to axis ***b****. The angles between these planes according to experimental data have the following values: ∠(011)* and (011‾)* = 82 °; ∠(012)* and (012‾)* = 50 °; ∠(013)* and (013‾)* = 33°; ∠(014)* and (014‾)* = 25°; ∠(032)* and (032‾) * = 106°. The graphical construction also shows that $d_{001} = c \sim 14.4$ Å.

Rhombic cell parameters measured directly on electronograms: ***a*** = 14.08 ± 0.03 Å, ***b*** = 13.22 ± 0.03 Å, ***c*** = 14.6 ± 0.09 Å. The result of parameter refinement by least squares method is as follows: ***a*** = 14.06 ± 0.03; ***b*** = 13.22 ± 0.02; ***c*** = 14.4 ± 0.12 Å (76 reflexes, $R_d = 0.88\%$). Angles between symmetric planes of type *(0kl)* and *(h0l)*, which experimentally fixed at goniometer rotation, are close to the values calculated. Among the *0kl* and *0 k0* type reflexes, only those with $k = 2n$ are represented. There are *h0l* reflexes with $h = 2n$. Reflexes of *hk0* type are characterized by even sum $h + k$, while *hkl* reflexes are characterized by even sum $h + k + l$. In *h00* and *00 l* series revealed existed only reflexes with even index values. Thus, urphoite space group *Ibca* was determined according to monocrystalline electron diffraction data. Calculated elementary cell volume V = 2676. When calculated on ideal mineral formula $UPO_4(OH){\cdot}nH_2O$, formular units number Z = 24. Calculated density D = 4.29 g/cm^3.

Powder X-ray pattern (Guinier camera, CuKα radiation) was obtained from sample where new mineral is most densely arranged (Dzhentuar deposit). It was impossible to completely get rid of impurities, so material for XRD was only fraction that enriched with new phosphate. Among impurities there are quartz, saléeite, pseudo-autunite, berthierine (that diagnosed by X-ray), and also halloysite, autunite, lepidocrocite, bassetite and pyrite. XRD decoding have been extremely complicated because of predominance of diffuse and unclear or clear but broad x-ray reflexes (Fig. 3.35), as well as impurity lines presence. Powder diffraction lines were indexed on the base of electron microdiffraction data (Table 3.5). On ring MD-patterns, broadening of reflexes is also noted because of texturing (Fig. 3.31g, e). In our case, mineral parameters are determined by method monocrystal (single crystal) diffraction, but not X-ray, and electron one.

Small size of new lamellar phosphate mineral formations, fine dispersion and impurities did not allow obtaining sufficient amount of pure, monomineral hand picking extraction for reliable X-ray diffraction characteristic of new mineral. Therefore, only electron microdiffraction results were used to determine parametric data. Judging by the lamellar-flake morphology of the new U^{4+}-phosphate microcrystals, we can assume a layered character of its structure. In this connection, the possibility of interlayer parameter ***c*** compression in instrument vacuum conditions is not excluding. That is due to removal of weakly bound water from the structure while preserving basis parameters unchanged, as it was noted, for example, in coconinoite study (Belova et al. 1993).

Table 3.5 Calculation results of powder x-ray Urphoite sample

	U^{4+} phosphate–Urphoite				Quartz ASTM 5-490		Pseudoautunite ASTM 18-1084		*Saléeite* ASTM 8-313		Berthierine[b]	
№	*I*	d meas.	d calc.	hkl	*I*	d	*I*	d	*I*	d	*I*	d
1	2 dif	9.3							10	9.94		
2	1	8.84							3	7.90		
3	2	7.2	7.2	002							10	7.02
4	1	6.9	7.03	200			6	6.78				
5	1 dif	6.59	6.61	020					3	6.61		
6	<4	6.3					10	6.20				
7	8 br	5.7–5.0	5.76–4.87	112, 022			2	5.10	8	5.01		
8	8	4.78	4.816	220			1	4.66			2	4.67
9	1	4.48							3 br	4.42		
10	1 br	4.41–4.37					1	4.38				
11	1	4.31[a]										
12	6	4.27			5	4.24					1/2	4.28
13	8	4.16	4.003	222			1	4.06				
14	2	3.72	3.74, 3.70	123, 321			3	3.72	1	3.71		
15	1	3.63	3.631, 3.60	132, 004								
16	10	3.52	3.515	400			3	3.57	9	3.52	10	3.52
17	5	3.38	3.372	114	10	3.34	5	3.38	3 br	3.34		
18	1	3.308	3.305	040, 411			10	3.25				
19	5 br	3.12÷3.06	3.103, 3.00	420, 042			8	2.95	3	3.12		
20	4	2.8	2.85	422			4	2.83	3	2.89	1	2.80
21	10 br	2.76÷2.71	2.77, 2.762	413, 242							4	2.68
22	4 br	2.55	2.547, 2.52	521, 404			2	2.65			9	2.52
23	2	2.479	2.444	152, 044	5	2.45	2	2.41	3	2.49		
24	1 br	2.32	2.343, 2.33	600, 116								
25	1	2.30	2.301, 2.28	244, 442	5	2.28	4	2.27				
26	2 br	2.25	2.256, 2.25	026, 532								

(continued)

Table 3.5 (continued)

	U^{4+} phosphate–Urphoite				Quartz ASTM 5-490		Pseudoautunite ASTM 18-1084		*Saléeite* ASTM 8-313		Berthierine[b]	
№	*I*	d meas.	d calc.	hkl	*I*	d	*I*	d	*I*	d	*I*	d
27	3 db	2.23	2.228, 2.21	602, 620	4	2.23	8	2.19	3br	2.23		
28	5 br	2.13	2.15, 2.112	161, 622	5	2.12	3	2.15			6	2.14
29	3 dif	2.08	2.10, 2.084	260, 136			3	2.106				
30	3 dr	2.00	2.018, 2.00	262, 444			8	2.06	2	2.00	1	2.01
31	1 dif	1.879	1.88, 1.867	064, 460			9	1.915	1	1.86		
32	1 dif	1.845	1.848	642								
33	4 dif	1.838	1.841	363								
34	5 br	1.814	1.816, 1.81	264, 172	9	1.813						
35	1	1.760	1.77, 1.76	732, 156			4	1.75	1	1.77	4	1.766
36	5 dif	1.753	1.758	800								
37	1	1.740	1.744, 1.74	208, 651								
38	1 br	1.702	1.707, 1.70	802, 820			4	1.71				
39	2 br	1.686	1.688, 1.68	644, 635			2	1.67	1	1.167		
40	1 br	1.635	1.631	181			4	1.65				
41	1 br	1.586	1.58, 1.579	266, 804					1	1.57		
42	4 br	1.550	1.552, 1.55	840, 381							7	1.555
43	2 br	1.528	1.532, 1.53	572, 716								
44	2 br	1.494	1.502	084								
45	2 br	1.438	1.44, 1.437	0.0.10, 574								
46	1	1.390	1.392	10.1.1								
47	2	1.386	1.386	826								
48	1 dif	1.345	1.35, 1.349	862, 491							1/2	1.347
49	1 dif	1.322	1.322	0.10.0			5	1.316			1/2	1.326
50	2 db	1.282	1.282	2.10.2			8	1.29				
51	2 dif	1.263	1.264	684								

[a]Line not diagnosed
[b]G.W. Brindley, G. Brown. Crystal structures of clay minerals & their X-Ray identification. Miner. Soc., 41 Queen's gate, London SW7 5HR, 1980
br broad; *dif* diffuse; *db* double

3.2.3.3 Location (Finds and Mineral Associations)

Finds locations. All discoveries of new phosphate was made in samples from oxidation zone of uranium deposits, and only it's association and formations character differ slightly. At Kanzhugan, Chu-Sarysu depression (Kazakhstan), this mineral was found as the main component of sand-clay-phosphate agglomerate. At Dzhentuar, Central Kyzyl-Kum (Uzbekistan), phosphate is found in form of very fine precipitates on the surface of lamellar uranyl minerals (uranium mica, coconinoite). At Koscheka the mineral is represented by nesting aggregations or interlayers in clay-quartz material of modified host rock; less frequently—by microspherulites on halloysite.

Mineral associations. In association with new tetravalent uranium phosphate, the most common are uranium mica (metaautunite, Fe-containing saléeite, bassetite); less frequently encountered are coconinoite (Al-uranyl phosphate), halloysite and layered silicates, and even more rarely—pyrite, quartz. Exclusively according to electron microscopy data, it has been established association with lepidocrocite, tyuyamunite, hydrogen uranium mica, ningyoite and native sulfur (Fig. 3.36).

Geological location. Mineral paragenesis shows that studied above U^{4+} phosphate was precipitated in complex process of hypergene mineral-formation, as sensitive reaction to narrow-local changes in redox environment in form of tetravalent uranium minerals that appears on the predominant background of hexavalent ones. This process localness is expressed in sharply subordinate amount and in dispersity of new mineral phase among obviously prevailing mass of uranium mica and other uranyl U^{6+}—minerals. Unlike other U^{4+}-phosphates, this mineral is found not only in secondary enrichment zone, but also in lowest horizons of own oxidative mineral-formation stages manifestation (on uranyl minerals surface). The presence of native sulfur indicates reductive geochemical conditions of mineral-formation environment.

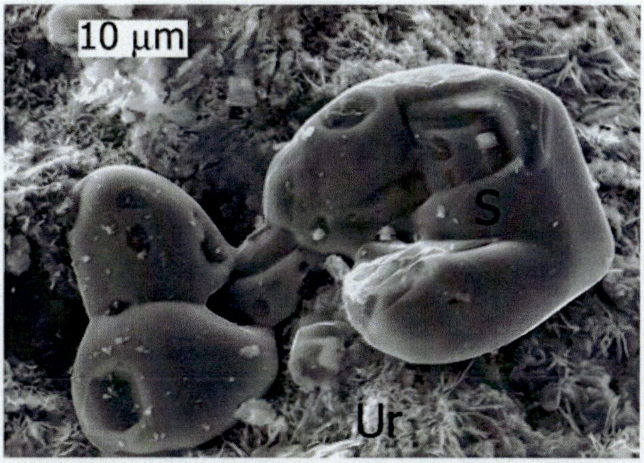

Fig. 3.36 Urphoite (Ur): association with native sulfur (S). SEM-image

3.2.3.4 Brief Historical Information of Paperwork for Urphoite

Material on our studied first mineral discovery in Kazakhstan was presented to editorial staff of Russian Academy Sciences Reports in December 1994, and published in late 1996. We did not give name to this mineral; it has been named "new phosphate of tetravalent uranium", but we did not doubt its individuality and novelty. Scientific work practice in mineralogy shows that in case of new minerals discovery the registration and subsequent publication period is fairly long. Therefore, we are convinced of the legality of this publication as priority (as notification). Subsequent this mineral finds at Uzbekistan deposits allowed us to determine optical characteristics, to clarify elementary cell symmetry.

So in 1997 we submitted application for new mineral discovery to Russian New Minerals Commission of VMO and further to NMNC IMA. The mineral is offered name "urphoite", corresponding to its composition: **ur**anous **pho**sphate—*urphoite*. NMNC IMA did not approve our presented uranium phosphate "urphoite". In Commission conclusion was dominated by generally fair remarks regarding unsatisfactory X-ray and composition data (only 2 analyses), which was due to lack of clean material. However, the IMA Commission, unfortunately, did not take into account the fact that in this case (when study high dispersed mineral) the advantage in diffraction studies belongs not to X-ray, but electron diffraction (SAED). In such cases of extremely small substance amount or its high dispersion state, electron-diffraction method (microdiffraction) is similar and equal to X-ray diffraction method in minerals diagnosis. Here SAED-method should be considered as method of monocrystalline diffraction study of object. Complex of electron microscopy analytical methods, which allows not only to analyze crystallographic characteristics, but also to determine composition of micron crystals, plays leading and decisive role. As examples—discovery and study of new minerals and mineral varieties such as: Vyacheslavite, Feroxyhyte, Akhtenskite, Ferrihydrite, Varlamoffite, Borodaevite, etc.

The Urphoite sample of L.N. Belova from Uzbekistan, Koscheka deposit, where green microcrystalline spherules of new phosphate are clearly distinguished on eroded light surface composed mainly of halloysite, was presented to collection of Fersman Mineralogical Museum RAS in Moscow.

Some Resume Currently, level of our knowledge on Lermontovite group minerals allows us to speak about the independence of each mineral species in it: Lermontovite, Vyacheslavite, and Urphoite. This conclusion is based on the individuality of structural parameters of each of them. The compositions similarity makes it possible to speak only presumably about polymorphy of these group minerals; for the final approval it is necessary both to carry out complete chemical analysis and to determine the structures. However, the extremely limited number of single U^{4+}-phosphates finds does not yet provide basis for this. Subsequent finds of this group minerals, which will be more significant in quantity, will make it possible to carry out detailed analysis of their chemical composition (with definition of hydroxyl groups and water content degree), as well as structural parameters refinement by X-ray or electron diffraction (PEDT) methods, to clarify remaining unresolved questions.

U^{4+}-Phosphates Color Characteristic t is known that there are numerous synthetic U^{4+}-phosphates none of which, however, is close in composition to minerals of the Lermontovite group with general formula $U(PO_4).nH_2O$. Due to the significant complexity of the study of Urphoite—one of this group mineral, the idea of obtaining similar synthetic compound arose. An attempt was made to synthesize such phosphate in laboratory conditions (analyst G.V. Lyubomilova). The obtained product of synthesis, composition $HU^{4+}{}_2(PO_4)_3{\cdot}7.8\ H_2O$, is indistinguishable from typical urphoite by color; it is X-ray- and electron-amorphous material, N = 1.720. It can be assumed that in future resynthesis at elevated temperatures and pressures will lead to production of crystallized material, as is the case with many uranium compounds.

The laboratory experiment on chemical synthesis of tetravalent uranium phosphate, although it did not lead to crystalline phase formation, was convincing confirmation of color characteristics of U^{4+}-phosphates. The green material coloring was extremely important to "remove" the question of uranium valence in green U^{4+}-phosphates. The question arose in connection with that by time of Vyacheslavite and Urphoite opening in uranium mineralogy there were known only *uranyl* compounds of green color (including U^{6+}-phosphates). There was an opinion that colored uranium minerals represent exclusively hexavalent uranium form (uranyl) $UO_2{}^{2+}$, in contrast to black and dark brown coloring of known U^{4+} minerals (uraninite, coffinite, brannerite). In that way, mineralogists' doubts about the coloration of tetravalent uranium phosphates are resolved. The laboratory synthesis of urphoite convincingly confirmed the color characteristic—green coloration of U^{4+}-phosphate.

Here it is important to note that recently discovered by Czech mineralogists group (J. Plášil and co-authors) new uranous minerals—sulphate behounekite $U(SO_4)_2{\cdot}4(H_2O)$ (Plášil et al. 2011) and arsenates vysokyite $U[AsO_2(OH)_2]_4{\cdot}4H_2O$ (Plášil et al. 2013b), štěpite $U(AsO_3OH)_2{\cdot}4H_2O$ (Plášil et al. 2013a)—also have green coloration: green, emerald green and light green consequently.

The diversity of synthetic tetravalent uranium phosphates suggests that the group of natural U^{4+}-phosphates will be systematically supplemented with new members.

3.2.4 Conclusions

The main results of natural U^{4+}-phosphates study by AEM methods are follows:

The phosphates discussed above are independent mineral species from the tetravalent uranium phosphate Family. Vyacheslavite is almost pure uranium phosphate with extremely low amounts of calcium; the ratio of mineral forming elements U: P = 1:1, but there is always slight predominance of phosphorus. In the Lermontovite composition calcium is even less, but about 10% thallium is present. The Urphoite composition in general is similar to formula $UPO_4OH \cdot nH_2O$, which is common for Lermontovite group; but here, it is some deviation: uranium content slightly increasing. In Ningyoite composition amount of calcium is equal or exceeds uranium amount. To date received data do not allow speaking definitely about the polymorphy of the Lermontovite group minerals or about their structures polytypity.

As a result of our crystallochemical researches, for the first time all currently available information on U^{4+}phosphates has been systematized. The mineral Family considered in above chapters is united by the common formation conditions: in the zone, where reduction stage of hypergenic redox processes manifests. This stage is typical both for hydrothermal uranium deposits weathering (intervals of uranium ores secondary enrichment) and for hydrogenic (infiltration) deposits. Association with uranous ore formations, so called uranium blacks, which "marking" zone of reducing conditions in mineral zonality, is very typical for U^{4+}-phosphates.

References

Atkin D, Basham IR, Bowles JFW (1983) Tristramite, a new calcium uranium phosphate of rhabdophane group. Miner Mag 47:393–396

Belova LN, Doinikova OA (2003a) Formation conditions of uranium minerals in oxidation zone of uranium deposits. Geol Ore Deposits 45(2):130–132

Belova LN, Doinikova OA (2003b) Conditions of uranium deposits redox zonality formation in zone of hypergenesis. New in Earth Sciences. Abstract from the International Conference, Moscow, MGGRU [in Russian]

Belova LN, Gorshkov AI, Ivanova OA (1978a) The first find of ningyoite in the USSR. Reports of the USSR Academy of Sciences, Doklady Akademii Nauk SSSR 238 (1):215–216 [in Russian]

Belova LN, Gorshkov AI, Ivanova OA (1978b) New data on ningyoite: Fe-containing ningyoite. Ibid 243(4):1022–1023. in Russian

Belova L.N., Ivanova O.A., Fedorov O.V., Ryzhov B.I., Gorshkov A.I., Lyubomilova G.V. About ningyoite ores of Bulgaria. In: Materials on geology of uranium, rare and rare-earth metals deposits. Inform Digest KNTS, Moscow (1979) 57, 63–68

Belova LN, Gorshkov AI, Ivanova OA, Sivtsov AV (1980) Nature of So-Called P-Bearing Coffinite. Reports USSR Acad. Sci. /Dokl. AN SSSR/. 255 (2), 428–430 [in Russian]

Belova LN, Gorshkov AI, Ivanova OA, Sivtsov AV, Boronikhin VA (1983) A new natural phosphate of U(IV). Reports of the USSR Academy of Sciences. Dokl Akad Nauk SSSR 273 (6):1460–1462. in Russian

Belova LN, Gorshkov AI, Ivanova OA, Sivtsov AV, Boronikhin VA (1984a) A new natural phosphate of U(IV). Int Geol Rev US 26(6):737–739

Belova LN, Gorshkov AI, Ivanova OA, Sivtsov AV, Lizorkina LI, Boronikhin Vyacheslavite VA (1984b) $U^{4+}(PO_4)(OH) \cdot nH_2O$, a new uranium phosphate. Zap Vses Mineral Obshch 113:360–365. in Russian

Belova LN, Gorshkov AI, Ivanova OA et al (1985) Ningyoite in the light of new experimental data. Proc USSR Acad Sci 3:101–109. in Russian

Belova LN, Gorshkov AI, Doinikova OA, Sivtsov AV, Zelenova O, Filippov IM, Drahomanov LV (1986) Ningyoite in Bulgaria. In: Crystal chemistry of minerals. XIII MMA Congress, Varna, vol 1. Bulgarian Academy of Sciences, Sofia, pp 773–779

Belova LN, Gorshkov AI, Doynikova OA, Sivtsov AV (1987) On analogy between tristramite and ningyoite. Izv Akad Nauk SSSR 11:136–137. in Russian

Belova LN, Gorshkov AI, Doynikova OA, Sivtsov AV, Mokhov AV, Trubkin NV (1993) New data about coconinoite. Doklady AN 329(6):772–775. in Russian

Belova LN, Gorshkov AI, Doinikova OA, Sivtsov AV, Trubkin NV (1996) Another one tetravalent uranium phosphate. Dokl RAS 349(3):361–363. in Russian

Belova LN, Gorshkov AI, Doynikova OA, Sivtsov AV (1998) A new group of minerals - phosphates U(IV). Dokl RAS 358(2):215–217. in Russian

Bokiy GB (1997) Systematics of natural silicates. Results of science and technology. Itogi Nauki i Tekhniki 31:292

Boronikhin VA, Tsepin AI (1980) Universal Program for calculating corrections and statistical processing of the results of current changes in quantitative X-ray spectral analysis ("PUMA"). In: Equipment and methods of X-ray analysis. L-d: Mashinostroenie Engineering 2, pp 204–217 [in Russian]

Bowles JFW, Morgan DJ (1984) The composition of rhabdophane. Miner Mag 48(1):146–148

Boyle DR, Littlejohn AL, Roberts AC et al (1981) Ningyoite in uranium deposits of south Central British-Columbia, first North American occurrence. Canad Miner 19(4):325–331

Burns PC, Finch RJ (1999) Wyartite: crystallographic evidence for the first pentavalent-uranium mineral. Am Miner 84(9):1456–1460

Buryanova EZ (1963) Determinant to minerals of uranium and thorium. Gosgeolizdat, 51 p

Doinikova OA (2003) Genetic crystal chemistry of mineral components in uranium blacks. Geochem Int 41(12):1214–1220

Doinikova OA (2007) Uranium deposits with a new phosphate type of blacks. Geol Ore Deposit 49 (1):80–86. https://doi.org/10.1134/S1075701507010047

Doinikova OA, Gorshkov AI, Belova LN et al (1993) Questions on systematics of phosphates rhabdophane group. Zapiski VMO 3:79–88. In Russian

Doynikova OA (2005) Dispersed uranium minerals of hypergenesis reducing zone: mineralogy and crystal-chemistry. Habilitation Thesis, Dissertation in Geology and Mineralogy, IGEM RAS, Moscow, p 277 [in Russian]

Doynikova O.A. (2012) Uranium mineralogy in hypergenesis reduction zone (according to electron microscopy data). Fizmatlit Moscow, 216 p [in Russian]

Doynikova OA, Belova LN, Gorshkov AI, Sivtsov AV (2003) Uranium blacks: questions of genesis and mineral composition. Geol Ore Deposit 45(6):514–530. in Russian

Doynikova OA, Tarasov NN, Mokhov AV (2014) New phosphatic type of uranium deposits in Russia. Doklady Earth Sci 457(Part 2):910–914. https://doi.org/10.1134/S1028334X14080030

Dusausoy Y, Germani NE, Podor R, Cuney M (1996) Low-temperature ordered phase of CaU $(PO_4)_2$: synthesis and crystal structure. Eur J Mineral 8(4):667–675

Dymkov YM (1985) Paragenesis of minerals of uranium-bearing veins. Nedra 207

Dymkov YM, Boitsov VE, Ivanova OA (1986) Ningyoite from hydrothermal veins in the Gorni Slavkov (ChSSR). Miner Zh 8(6):34–43. in Russian

Getseva RV, Savelieva KT (1957) Guidelines for the determination of uranium minerals. Gosgeolizdat, Moscow, p 260. in Russian

Steciuk G, Ghazisaeed S, Kiefer B, Plášil J (June 2019) Crystal structure of vyacheslavite, U(PO4)(OH), solved from natural nanocrystal: a precession electron diffraction tomography (PEDT) study and DFT calculations. RSC Adv 9(34):19657–19661. https://doi.org/10.1039/c9ra03694f

Kajitani K. (1970) A geochemical study on the genesis of ningyoite, the special calcium uranous phosphate mineral. Econ Geol. 65: 470–480.

Kislyakov YM, Shchetochkin VN (2000) Hydrogenic ore formation. Geoinformark, Moscow 608 [in Russian]

Kniewald G, Palinkas LA, Bermanec V (2002) Hydrogeochemical and thermodynamic controls on the REDOX speciation of uranium in aqueous solutions and U-ores. 11 JAGOD Quadrennial Simposium and Geocongress, Winhoek, pp 432–433

Kucha H, Weiczorek A (1980) $Ca_{1-x}Th_{1-x}RE_{2x}$ $[PO_4]_2 \cdot 2H_2O$, a new mineral from Lower Silesia, Poland. Miner Polonica 11:123–136

Lisitsin AK (1975) Hydrogeochemistry of ore formation. Nedra, Moscow, p 247. in Russian

Melkov VG, Pukhalsky LC (1957) Searches for uranium deposits. Gosgeoltehizdat, 220 [in Russian]

Melkov VG, Belova LN, Gorshkov AI, Ivanova OA, Sivtsov AV (1983) New data on lermontovite. Mineral J 5(1):82–86. in Russian

Muto T (1962) The precipitation environment of ningyoite. Miner J 3(5–6):306–309

Muto T (1965) Thermochemical stability of ningyoite. Mineral J (Jpn) 4:245–274

Muto T, Meyrowitz R, Pommer AM, Murano T (1959) Ningyoite – a new uranous phosphate mineral from Japan. Am Mineral 44(5–6):633–639

Perelman AI (1968) Geochemistry of epigenetic processes (hypergenesis zone), Nedra 331 [in Russian]

Petrovskaya NV, Novgorodova MI et al (1976) Informations. USSR Academy Sci (3):67–71. in Russian

Plášil J, Fejfarová K, Novák M, Dušek M, Škoda R, Hloušek J, Čejka J, Majzlan J, Sejkora J, Machovič V, Talla D (2011) Běhounekite, $U(SO_4)_2(H_2O)_4$, from Jáchymov (St Joachimsthal), Czech Republic: the first natural U^{4+} sulphate. Mineral Mag 75:2739–2753

Plášil J, Fejfarová K, Hloušek J, Škoda R, Novák M, Sejkora J, Čejka J, Dušek M, Veselovský F, Ondruš P, Majzlan J, Mrázek Z (2013a) Štěpite, $U(AsO_3OH)_2{\cdot}4H_2O$, from Jáchymov, Czech Republic: the first natural arsenate of tetravalent uranium. Mineral Mag 77:137–152

Plášil J, Hloušek J, Škoda R, Novák M, Sejkora J, Čejka J, Veselovský F, Majzlan J (2013b) Vysokyite, $U^{4+}[AsO_2(OH)_2]4{\cdot}4H_2O$, a new mineral from Jáchymov, Czech Republic. Mineral Mag 77:3055–3066

Proceedings of the International Conference on the atomic energy peaceful use (1958) Geneva (1955) Gosnauchtekhizdat 6:966 [in Russian]

Ryzhov BI, Korobenko IR, Doinikova OA (1985) Materials on geology of uranium, rare and rare-earth metals deposits. Inform Digest KNTS (95):52–57. in Russian

Scharm B (1993) Některé vzácné minerály provázejici uranové zrudněni v Severočeská Křidě. Bull Mineral Petrogr (Praha):45–48

Scharm B, Hofreiter V (1978) Stručný přehled dosavadních poznatků z mineralogie uranových ložisek v severočeské křídě (strážský blok). Geol Hydrometalurgie uranu 2(2):3–26. in Czech

Scharm B, Burda J, Sulovsfry P, Scharmova M (1980) Remarkable mineral assemblage on uranium deposits in northern Bohemia (the Strazh block). Chasopis pro Mineral Geol 25(2):113–124

Scharm B, Scharmova M, Kundrat M (1994) Crandalite group minerals in the uranium ore district of Northern Bohemia (Czech Republic). Věst Čes Geol Ust (Praha) 69(1):79–85

Scharmova M, Scharm B (1994) Rhabdophane group minerals in the uranium ore district of northern Bohemia (Czech Republic). J Czech Geol Soc 39(4):267–280

Scharmová M, Půlpán T, Sulovský P, Vjačeslavit SB (1990) $U^{4+}PO_4{\cdot}OH{\cdot}H_2O$, ze severočeské křídy – první výskyt v Československu Stráž p. Ralskem/Membrany. Ecol Geol Anal 14 (3):81–87

Scharmova M, Scharm B, Rutšek J (1993) Rabdofan-(Nd), Churchit-(Y) a Langizit na lozisku Stráž v Severočeske Křidě, Sbor. V Mineralogický cyclický seminař (Ústí na Labem), pp 97–99

Sidorenko GA (1960) X-ray-graphical determinant of uranium minerals, Energoizdat, 192 p

Sidorenko GA, Gorobets BS, Dubinchuk VT (1986) Modern methods of mineralogical analysis of uranium ores. Energoatomizdat. Moscow. 184 [in Russian]

Soboleva MV, Pudovkina IA (1957) Uranium minerals. Gosgeolizdat, 408 p

Teterin YA, Kulakov VM, Baev AS, Nevzorov NB, Melnikov IV, Streltsov VA, Mashirov LG, Suglobov DN, Zelenkov AG (1981) A study of synthetic and natural uranium oxides by X ray photoelectron spectroscopy. Phys Chem Miner 7(4):151–158

Vasiliev EK, Kashaeva GM (1974) Ushchapovskaya 3. F. Roentgen-metric determinant of minerals (phosphate class). Nauka, 206 p

Chapter 4
Some Aspects of Uranous (U^{4+}) Mineralogy in the Light of Crystallochemical ATEM Data

4.1 Genetic Crystallochemistry of Uranous Minerals

The section is devoted to identification of geochemical conditions of the uranium blacks' formation on the basis of determining the crystallization conditions of various uranium minerals that make up these blacks. Based on the ATEM data, the geochemical features of U^{4+}-minerals are determined. It is further shown that features the elemental composition and isomorphism permissible in the mineral structure can, in certain cases, uniquely characterize the conditions of its crystallization. It is considered how the geochemical features of minerals coffinite and ningyoite indicate conditions of uranium black ore formation.

The terms "*genetic crystallochemistry*" (Frank-Kamenetsky 1978), and "*crystallochemical geochemistry*" (Belov 1982) proposed by Russian crystallographers refers to the same "science about the structure of crystalline substances, which gives insight into processes of natural compounds formation. Since the publication "Essays on structural Mineralogy" in 1976 N.V. Belov discussed geochemical processes connected with minerals and rocks formation, based on mineral structural data. As defined by acad. N.V. Belov, crystallochemical geochemistry considers "problems of the structure of crystalline matter ... and problems of processes in which these substances arise". The focus of this science is "the relationship between the formation of crystalline matter (minerals) in nature, the transformation of minerals in the processes of epigenesis, metamorphism, and of various external influences ...".

"The problems of genetic crystallochemistry of minerals opening up new possibilities for revealing the links between the crystal structure of minerals and the processes of their formation ... "and "relationship between the real structure of mineral and its geological past" (Frank-Kamenetsky 1985). Since the 2000s the problems of genetic crystallochemistry been raised to new level of mineral structure research (Urusov 2001).

O. A. Doynikova, *Uranous Mineralogy of Hypergene Reduction Region*, Springer Mineralogy, https://doi.org/10.1007/978-3-030-67183-9_4

Since the main focus of all geological research (including mineralogy, crystallochemistry, and many other methods of substance studying) is, ultimately, the clarification of ore formation conditions, it was interesting to attempt to obtain genetic information through available crystallochemical data. The author of this book has attempted to obtain genetic information through available crystallochemical data. The *geochemical conditions of crystallization* of disperse U^{4+}-minerals can be judged by their elemental composition based on the researches of A.I. Perelman (1968). The reasoning was based on the specificity of U^{4+} migration in aqueous solutions and regularities of uranium concentration in hypergenesis region. In this case, the dispersity of mineral substance is extremely important, because this fact allows us to speak about the closeness in time of processes migration and crystallization, and even, with some assumption, about their simultaneity—in the geological sense. From this point of view, the available information on the composition of uranium black' mineral components—coffinites and ningyoites from various ore samples—was analyzed in the genetic aspect.

4.1.1 Obtaining Genetic Information through Available Crystallochemical Data

The data on uranium minerals crystallochemistry reflect the peculiarities of mineral formation and allow determining the regularities of migration and deposition of ore components in the form of minerals within the hypergenesis region. Crystallochemical peculiarities of mineral—composition and character of structural isomorphic transformations—serve as indicators of geochemical conditions of mineral formation. Let's see how real minerals that make up uranium blacks (coffinite and ningyoite) indicate the geochemical environment conditions through their elemental composition.

According to A.I. Perelman, uranium migration in the hypergenesis region can take place as in oxygen waters (when U^{6+}) and so in oxygen-free gley ones (when U^{4+}) (Hydrogenic Uranium Deposits 1980; Perelman 1968, 1989). Each type of waters (by Eh-pH combination) has its own specific association of migrating together *paragenous* chemical elements; and there is association of elements that migration here banned. And although uranium transport in natural waters is carried out mainly in hexavalent form, A.I. Perelman supposes the possibility of its migration to short distances in tetravalent form, under certain conditions.

Identification of conditions formation of uranium minerals and, consequently, of uranium ores composed by them, consists in the analysis of types of geochemical environments. It is possible to determine the mineral formation geochemical conditions, by which created uranous ores, based on research devoted to uranium geochemistry in hypergenesis region (Perelman 1968, 1989; Perelman and Kasimov 1999). It can be evaluated by peculiarities of elements migration and regularities of uranium concentration during hypergenesis that have been identify earlier. Having set of elements fixed in mineral composition and analyzing the conditions of simultaneous mobility of these elements, the *geochemical conditions of crystallization* of U^{4+}-minerals can be determine. Consequently, it is possible to determine type of geochemical environment according to A.I. Perelman. This is done through the proposed by him Systematics of *geochemical barriers,* i.e. *types of geochemical* environments in hypergenesis region. The assigned types of environments are given in form of table in various Perelman's monographs, one of the last (Perelman and Kasimov 1999). /Formulated by A.I. Perelman, *geochemical barriers* are those parts of Earth's crust, where at a short distance there is sharp decrease in intensity migration of chemical elements and, consequently, their concentration take place./ The types of uranium concentration conditions on reducing barriers in hypergenesis region have been considered in detail earlier as special case (Hydrogenic Uranium Deposits 1980).

The elements included in dispersed U^{4+}-mineral are considered this as paragenous ones, simultaneously with uranium (U^{4+}) coexisting (and mobile) in ore-forming aqueous solutions. Based on A.I. Perelman's classification of geochemical environments (Hydrogenic Uranium Deposits 1980), the analysis of conditions under which other composition elements are mobile in solutions simultaneously with U^{4+}, allows make a conclusion on the geochemical environment of mineral formation.

Determining the composition of micron crystal, we get set of elements that form structure and enter the mineral in form of impurities. Having chosen in the preparation the particle (without signs of leaching/dissolution), diffraction pattern of which confirms its monomineral nature, we determine its composition. The analysis of monomineral particle determines set of elements that coexisted in solution during this mineral crystallization and, therefore, were included in its composition. As a result, we define *paragenous* elements in this way.

Paragenous elements (both basic and impurity) coexisting with uranium in mineral composition indicate their simultaneous water migration. Since uranium blacks are formed mainly as result of hydrogenic mineral-formation in sedimentary deposits we talk about water migration of elements that is considered here. Strictly speaking, the impurity elements can be considered to be paragenous (along with the main mineral-forming elements) only in case of their isomorphic entry into mineral structure. However, impurities captured in crystallization process, and sorbed by surface, characterize mineral formation conditions to a certain extent too.

It is possible to determine the geochemical conditions of mineral formation that created uranium blacks based on research devoted to uranium geochemistry in hypergenesis region (Perelman 1989; Perelman and Kasimov 1999). Having set of elements fixed in mineral composition and analyzing the conditions of simultaneous mobility of these elements, according to A.I. Perelman, it is possible to determine geochemical type of environment. It should be noted that identification the conditions of elements water-migration and conditions of this elements deposition (as minerals) is correct only in case of simultaneous occurrence of these processes. The indicator of such geological simultaneity of these processes is dispersity of mineral substance. The micron dimension of minerals indicates the relative short-term geological duration of crystallization process.

It is micron (not more *n*· μm) size of crystals coffinite and ningyoite present in uranium blacks' composition that makes it possible to determine the paragenous elements. The analysis of the conditions under which these elements composing uranium minerals are mobil makes it possible to draw a conclusion about the geochemical environment of uranium ore formation. The relationship between the mineralization of uranium black ores and geochemical barriers is discussed below.

4.1.2 Genetic Analysis of Crystallochemical Data

When consider, for example, the widely spread in uranium blacks mineral—U^{4+}-silicate coffinite $U^{4+}(SiO_4)_{1-x}(OH)_{4x}$. Presence of impurity P, As, Ca, Ti, V and Fe in composition of various *coffinites* (see Chap. 2) allows to characterize geochemical environment as reducing *gley* (without H_2S) setting (medium) (Doynikova 2012), as only here joint migration of each of listed elements, simultaneously with uranium and silicon, is possible (Perelman 1989). This set of paragenous elements indicates the environment of geochemical barrier in which tetravalent uranium silicate is deposited. This situation corresponds to number of reducing geochemical barriers of gley and sulfide (H_2S) type.

Clarify the geochemical conditions of *ningyoite* formation help established earlier isomorphic impurities in its composition. So, for Fe-containing ningyoite, U^{4+} and divalent iron Fe^{2+} (possibly isomorphic to calcium) into the structure of mineral indicates a *weakly gley* reducing transport medium. A set of paragenous elements (Ca, U^{4+}, P, Fe^{2+}) indicates uranium precipitating sulfide barriers and gley type of waters where iron Fe^{2+} migration is allowed. According A.I. Perelman (1989) such barriers may be sulfide formations in ore-bearing stratum. However, recent data point to the predominantly biogenic nature of the barrier that precipitates uranium; leaving for sulfides only the role of closely associated mineralization. Other ningyoite impurities of As, V, Y, S, Ce, Ti captured during crystallization also indicate the predominant *weak gley* environment of its formation in hypergene conditions. As judged by isomorphic impurities (Fe, Y, As, S, V and Ce) that found in various ningyoites, the prevailing environment of ningyoite crystallization in hypergene conditions is *weakly gley* too (Doynikova 2012). It is in this environment that the water migration of the above listed elements is allowed simultaneously with uranium U^{4+} and phosphorus.

The analysis of uranium ores mineral associations allows to judge about the nature of precipitating barriers. As noted above, ningyoite–pyrite interbedding is indicated a sulfide precipitating barrier, where crystallization of phosphate occurs in a sharply reducing hydrogen sulfide environment (samples Momino, Bulgaria). Sulfide mineralization or carbonaceous organic matter can serve as reducing barriers in weakly reducing conditions. The association of Fe-containing ningyoite with spherical zincite (?) particles, and with pyrite absence in ore composition, points to alkaline environment of crystallization (Navysen, Bulgaria).

These examples show how the reducing geochemical environment (gley or hydrogen sulfide), typical for uranium blacks formation and deposition in hypergenesis region, is fixed (reflected) in the composition of its mineral components. However, using of elements set that paragenous with uranium (in composition of uranous crystalline phases—silicate or phosphate) allows only clarification of mineral formation conditions. More precise determination of geochemical environment type at any deposit requires involvement of detailed geological and mineralogical characteristics on a concrete ore occurrence. For example, according to (Hydrogenic Uranium Deposits 1980), oxygen character of waters and, accordingly, oxygen mineral zonality types will be indicated by warm color tones of formed oxidation zone (yellow-ginger-reddish-fulvous gamma). Gley waters and types of gley mineral zonality will be indicated by cold tones of coloring (white-bluish-grey-green gamma).

4.1.3 Genetic Conclusions from Uranous Minerals' Crystallochemical Data

As we can see above, in some cases, set of paragenic elements fixed in single crystal allows determining the conditions of mineral formation in uranium blacks with a high degree of probability. Besides that, elemental minerals composition can also serve as criterion for isomorphic substitutions character, as will be shown below when considering isomorphism in rhabdophane-like structures.

Thus, analysis of geochemical environments types, according to A.I. Perelman's classification (Perelman 1989), by set of established paragenous elements, allows us to draw conclusions about geochemical situation in which industrial clusters of uranium black were formed. As a result, crystallochemical data on the mineral serve as basis for genetic analysis.

Dispersity of uranous minerals (dimension ~ n· μm) in ores testifies rather short-duration crystallization, which indicates relatively narrow time interval in which deposition conditions (environment) and solutions composition remained unchanged. In this case dispersity that inherent for blacks' formations is considered as indicator of mineral formation conditions. The clear confinement of U^{4+}-minerals formations to boundary of reducing and oxidizing medium allows considering such mineral formation zone, in its essence, as geochemical barrier.

That can be considered as illustration of Perelman's assertion: "*decreasing of space occupied by environment leads to conversion of quantity into quality: the environment becomes a barrier (and vice versa)*". In this case, spaces volume in which uranous ore is forming representing by self the geochemical barrier that precipitated uranium. Having established geochemical environment for crystallization of uranium minerals (in blacks' ore), we thus determine the type of precipitating barrier for uranium. In particular case, ningyoite U-ores of Kosachinoye deposit are considered as a geochemical phosphate barrier accumulating beryllium (see Chap. 2).

Comparing the main uranium migration forms in aqueous solutions, as well as on the basis of small solubility values of tetravalent uranium compounds, A.I. Perelman assumed that "conditions of stable existence of tetravalent uranium are not favorable for its migration in underground waters" (Hydrogenic Uranium Deposits 1980). Different finds of exclusively micron-sized crystals ningyoites and coffinites prove the validity of this assumption. It is interesting to note that conditions for the short stable existence of U^{4+} in solution are logically explained by the biochemical nature of ore-accumulation barrier, if we accepting biogenic formation of uranium deposits (Vinichenko 2004). Here implying the increased concentration of organic-mineral uranium compounds in solution and, at the same time, a sharply increased number of anaerobic bacteria, which takes place in reducing zone of redox biochemical barrier. If we assuming biogenic formation, it also possible to resolve question on the reason for extremely small, exclusively micron size of crystals (ningyoite, coffinite) and other uranous mineral formations (uraninite) in blacks' ores; that's discussed hereinafter.

The results of crystallochemical study of uranous mineralization (U blacks) by ATEM methods allow us to make following conclusions: • established crystallochemical features of uranium minerals microcrystals are indicators of geochemical environment and allow determining qualitatively conditions of ore mineral formation process; • paragenic elements, recorded in these minerals composition, indicate their simultaneity presence in aquatic environment and, thus, characterize the U-bearing solutions and medium of mineral formation.

4.2 Questions of Isomorphism and Systematics of Phosphates in Rhabdophane Mineral Group

Many years experience in studying ningyoite from different deposits testifies to existence of variations in its cationic composition. Taking into account similarity of ningyoite crystal structure and REE-phosphate, rhabdophane, problems of isomorphism in ningyoite composition have been consider on basis of composition analysis of all this group's minerals. Systematics of Rhabdophane group's phosphates was specified by analyzing of both known literary data and our own researches data on this group's minerals (Doinikova et al. 1993; Doynikova 2012). ATEM methods were used to study reference samples of Th-phosphates brockite and smirnovskite from Fersman Mineralogical Museum RAS in Moscow. It is shown that all minerals in Rhabdophane group are characterized by rhabdophane-like structure; the issues of isomorphic transformations in this mineral group are analyzed on example of ningyoite as its typical representative member.

4.2.1 Mineral Composition of Rhabdophane Group

According to modern reference data, Rhabdophane group of rare-earth phosphate, includes its own 4 varieties, differing in prevailing of REE-component (Ce, La, Nd, Y), as so Brockite, Tristramite and Unnamed mineral phase. Group formula: (REE, Ca, Th) $(PO_4) \cdot H_2O$ (https://www.mindat.org/min-29326.html). M. Fleischer (1990) includes here both REE-phosphates and four other minerals: brockite, grayite, ningyoite and tristramite. General formula for this group members is given by M. Fleischer as $XZO_4 \cdot 1\text{-}2H_2O$, where X = Ca, Ce, La, Fe^{3+}, Pb, Th and Y (since 2012); Z = P, S. In Russian reference books, besides REE phosphates, Brockite and Ningyoite included group (Mineralogical Tables 1981), and previously Smirnovskite been included (Vasilyev and Kashaeva 1974). Due to reference data ambiguity, it became necessary to clarify thc issuc of Rhabdophane group mineral composition on the basis of APEM crystallochemical data received by author.

The mineral composition of the group was analyzed by the author and corrections were made. Our data allowed us to refine and supplement list of X-cations indicated by M. Fleisher. The presence of Pb cations appears to be questionable in this list (results of Smirnovskite study see below). Previously, U wasn't on the cation list. In tristramite composition bivalent iron Fe^{2+} should be indicated, which follows from determining of geochemical environment formation of its analogue ningyoite (discovered earlier)—this environment is reductive. However, to date, valence of Fe^{3+} is erroneously indicated (https://www.mindat.org/min-4025.html), without taking into account its structural position within this mineral and the geochemistry of mineral formation. On our results of ningyoite study it should be indicated as Fe^{2+}.

According to our data, the cationic composition in generalized formula of Rhabdophane minerals group should be represented as follows: X = REE, Th, Ca, Fe^{2+}, U; where REE = Ce, La, Nd and Y. Our data proves that anion composition of common formula includes P and S (Chap. 3, sect.1).

Rhabdophane is aqueous Ce-phosphate of hexagonal syngony; it is the most widespread member in this group, described in detail in various reference sources. Its sp. group $P6_222$; elementary cell parameters of its natural mineral sample are as follows: $a = 6.98$, $c = 6.39$ Å (ASTM 12–277).

According to MMA, currently there are four types of rhabdophane, differing in the prevailing REE-component: cerium (Ce), lanthanum (La), neodymium (Nd) and yttrium (Y). Yttrium variety added since 2012. According to Russian mineralogist P.M. Kartashov (oral communication), these minerals are slightly different only in shades of color: rhabdophane (Ce)—yellowish, rhabdophane (Nd)—pinkish (because it is always associated with red-colored mineral associations); rhabdophane (La)—is considered to be colorless; rhabdophane (Y)—yellowish white to yellowish brown. Color is difficult to determine due to extremely small size of its formations.

Ningyoite is aqueous Ca-U-phosphate, $CaU(PO_4)_2 \cdot 1\text{–}2\ H_2O$, described in detail in the previous chapter. Mineral was discovered in Japan and approved in 1959. Until 1985 in all publications, including our first articles, Ningyoite was considered as rhombic, pseudo-hexagonal with cell parameters: a = 6,78 Å, b = 12,1 Å, *c* = 6,38 Å (https://www.mindat.org/min-2906.html). Conclusion Japanese authors about structures similarity of ningyoite and rhabdophane, which was previously assumed by the similarity of powder X-ray films, is confirmed by our SAED study of reference sample Japanese ningyoite. Diffraction study of ningyoite microcrystals from various samples proved that the real symmetry of ningyoite structure is hexagonal (see Section I). The previously selected face-centered (C) unit cell with orthogonal crystallographic axes is actually orthohexagonal (when b = a · √3), which often happens when choosing a unit cell in hexagonal lattice; it is well known in crystallography. Detailed SAED study of this phosphate made it possible for the first time establishes true hexagonal nature of its crystal lattice. Elementary cell parameters of natural ningyoite, established on basis of SAED data: $a = 6.87$, $c = 6.40$ Å; sp. gr. $P6_222$ (Belova et al. 1985). A variety of this mineral with the rare earth elements in composition, mainly yttrium (Y) group, was found (Ivanova et al. 1985).

Tristramite, rhabdophane-like Ca-U phosphate, was discovered in England, UK, in 1982 and approved by IMA (CNMNC) as new mineral with formula $(Ca,U^{4+},Fe^{3+})(PO_4,SO_4)\cdot 2H_2O$; $a = 6.91$ Å, $c = 6.42$ Å (Atkin et al. 1983). When approving this new mineral species, hexagonal syngony was considered the main its difference from Ningyoite, which at that time was considered to be orthorhombic. Later, the mono-crystal SAED method showed us hexagonal ningyoite symmetry (Belova et al. 1985). This is established both in our samples and in the reference ningyoite sample from Japan. Therefore, according to Russian mineralogists (Belova et al. 1987), Tristramite is analogue of Ningyoite, previously discovered in Japan. The identity of these minerals is confirmed by redefined parameters of ningyoite hexagonal cell.

Discoverers Tristramite noted that it is impregnated with Goethite. Probably, it is this mineral association that led the authors to erroneously recalculate all the established iron to Fe^{3+}. This may have led to more than strange conclusion about the entry of trivalent (oxidized!) iron Fe^{3+} into cation part of formula, which is unacceptable here (in the mineral) simultaneously with reduced uranium form U^{4+}. As a consequence, all reference data incorrectly shows the iron valence Fe^{3+} in the formula (https://www.mindat.org/min-4025.html). The set of crystal chemical data indicates divalent state of iron entering into in mineral structure, presumably isomorphically to Ca. So author decides that Tristramite is not individual (self dependent) mineral, it is variety of earlier discovered Ningyoite.

Brockite is aqueous phosphate with formula $(Ca,Th)PO_4\cdot H_2O$ and X-ray similar to Rhabdophane, discovered in Colorado, USA (Fisher and Meyrowitz 1962). It was found in form of massive reddish aggregates of cryptocrystalline substance or as earthy yellow crusts and concretions in altered granite rocks. In concentrations, usual size of individual grains is near 20 μm. This fine the substance dispersion made it extremely difficult obtaining pure mineral fraction for testing; purity of sample was determined by X-ray. Semiquantitative analysis (from sample 100 mg) gave the main components ratio in composition CaO: ThO_2: P_2O_5 = 1:1:1. The low crystallinity of studied material is manifested so that even the best X-ray line is wide enough. By discoverers, indication was carried out by analogy with rhabdophane on

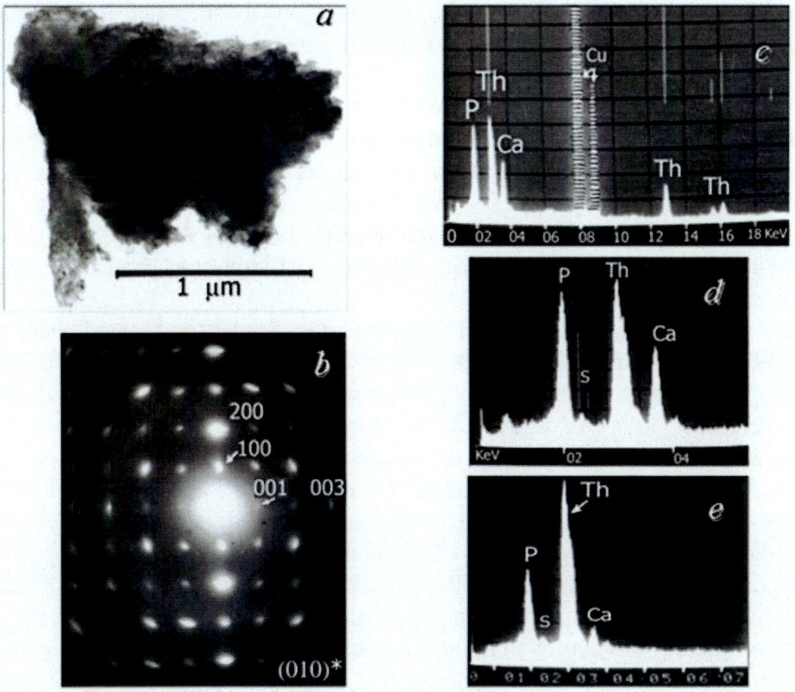

Fig. 4.1 Brockite. (**a**) TEM- image brockite particle; (**b**) point SAED pattern, plane (010)*; (**c**) composition ED- spectrum and its fragment (**d**) Cu peaks from preparation; (**e**) spectrum with reduced content Ca and P in composition from brockite particles from the same preparation. Sample from Fersman Mineralogical Museum, Moscow; No 76057

basis of hexagonal cell and identity of spatial group; cell parameters $a = 6.98$ Å, c = 6.40 Å.

Studied by us Brockite sample from Fersman Mineralogical Museum RAS, Moscow, (No. 76057) is characterized by hexagonal rhabdophane-type elementary cell with parameters: a = 7.03, c = 6.40 ± 0.03 Å (Doinikova et al. 1993). Composition of museum Brockite is represented in EDS spectrum by peaks P, Th, Ca (main), small S peak, and at background level—REE and Fe (Fig. 4.1). It follows from peaks relation of this phosphate spectrum that in cationic part of this mineral Th ≈ Ca, in anionic one P > > S.

Grayite is thorium phosphate, $(Th,Pb,Ca)(PO_4) \cdot H_2O$, structurally similar to rhabdophane according to X-rays, found (Bowie 1957) in Southern Rhodesia (Zimbabwe) and later in Wyoming and Colorado, USA (Dooley and Hathaway 1961). In these two known finds, mineral was in powder form or in form of very thin micron fraction, which did not allow discoverers studied in detail optical properties and determine its composition more accurately than semi-quantitative spectral method. Insufficiently precise composition characterization, such as "... little Ca, Pb; little U and TR..." led to fact that in M. Fleischer's reference book (1990), Pb is in list of mineral-forming cations. Obviously, such data are not sufficient to justify Pb entering to cationic part of formula. This is indirectly indicated by variable lead content of Smirnovskite particles (see below).

When compared Grayite and later discovered but more thoroughly studied Brockite, F. Fisher and R. Meyrowitz (Fisher and Meyrowitz 1962) is found difference in their belonging to different syngony, i.e. in pseudo- and true hexagonality of their space lattice. The report on grayite pseudo-hexagonality (Dooley and Hathaway 1961) is based on the presence in X-ray image of two diffraction lines that not indicated in hexagonal cell (4.54 Å and 1.81 Å). It seems to be insufficiently convincing, since probability of impurities presence in analyzed powder material is not excluded.

E.I. Semenov (1963, p. 155) considers Grayite as undoubted analogue of Smirnovskite. Smirnovskite, in turn, he considers as Rhabdofan's Th-analogue, although he does not identify Smirnovskite with the Broker. Following E.I. Semenov, author of this work considering that Grayite is not a new mineral species, but it is powder variety of Th-phosphate from Rhabdophane group. The absence of more detailed later publications about Grayite allows author to think so.

Smirnovskite was found in Transbaikalia and was first declared as Th-silicophosphate (Grigoriev and Dolomanova 1957), so it was not approved as independent individual mineral (Fleischer 1962). Later publication on Smirnovskite clarifies its composition as thorium phosphate (Dolomanova and Borisovsky 1979), and this mineral was approved as questionable "possibly the same as Brockite" with formula $(Th,Ca)PO_4{\cdot}nH_2O$. To prove that Smirnovskite is Brockite's analogue, reference Smirnovskite sample (№ 59990) from Vernadsky Mineralogical Museum in Moscow was studied by us.

In studied sample, mineral is represented in TEM mainly by amorphous particles of thorium phosphate (Fig. 4.2a) with low Ca content and very low Pb and Fe admixtures, at background level (Fig. 4.2d, e). Lead, as a rule, clearly identified only in amorphous, non-crystalline, particles; therefore, entering of Pb in composition of mineral (its cation part) is doubtful and requires more detailed verification.

The lack of diffraction is explained by metamict state, characteristic for minerals containing radioactive elements (Th). At the same time, particles that diffract sufficiently well under electron beam were also found. Point and ring electronograms were obtained (Fig. 4.2b, c). By intensity ratio and geometry of reflexes location, they are similar to the diffraction patterns of well-studied ningyoites. The parameters of hexagonal Smirnovskite cell were calculated: $a = 7.12 \pm 0.04$, $c = 6.47 \pm 0.04$ Å (Doinikova et al. 1993). Therefore, Th-Ca-phosphate Smirnovskite is structurally similar to ningyoite and is mineral of the Rhabdophane group.

The author conducted research allowed to prove that Smirnovskite from Transbaikalia practically is analogue of Brockite discovered later in Colorado. We should pay tribute to high professionalism of Russian mineralogists, who, in fact, have priority in discovery of this mineral species (Grigoriev and Dolomanova 1957). Only weak previous study of Smirnovskite, due to limited instrumentation and analytical equipment in those years, caused doubt about individuality of this mineral.

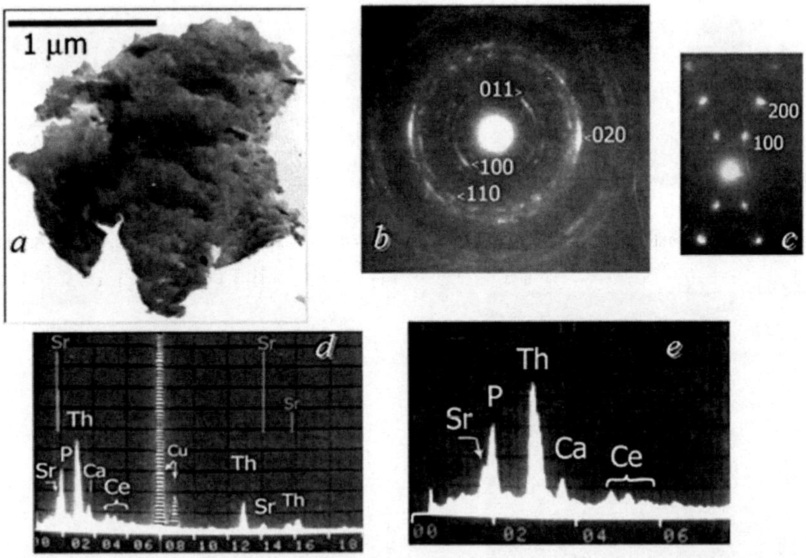

Fig. 4.2 Smirnovskite. Typical TEM-image of particles (**a**), rare ring SAED pattern (**b**); very rare point MD pattern (**c**); ED- spectrum of composition (**d**) and its fragment (**e**) shaded Cu peaks—from preparation. Sample from Fersman Mineralogical Museum, Moscow; No 59990

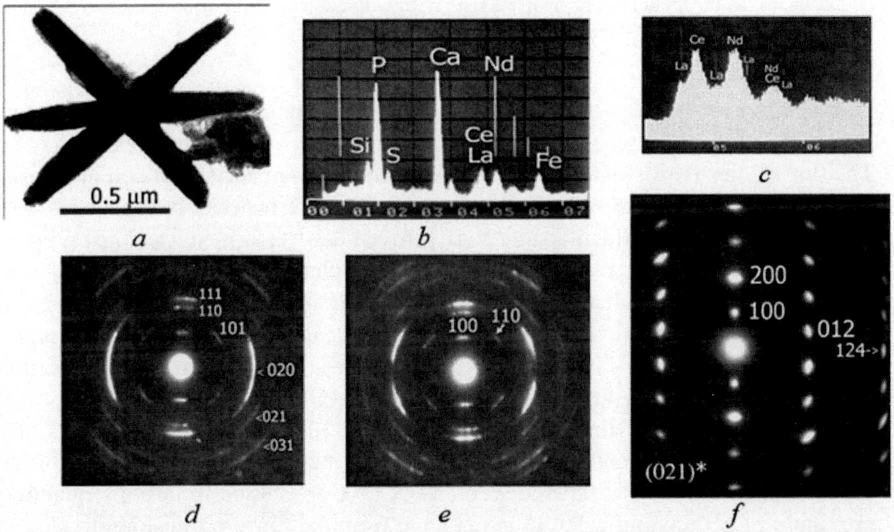

Fig. 4.3 Ca—rhabdophane variety (from apatite inclusions, Aldan craton, sample of N.V. Guliy). Microcrystal intergrowth image TEM (**a**); ED- spectrum (**b**) and its fragment (**c**); ring textured (**d**, **e**) and point (**f**) SAED patterns

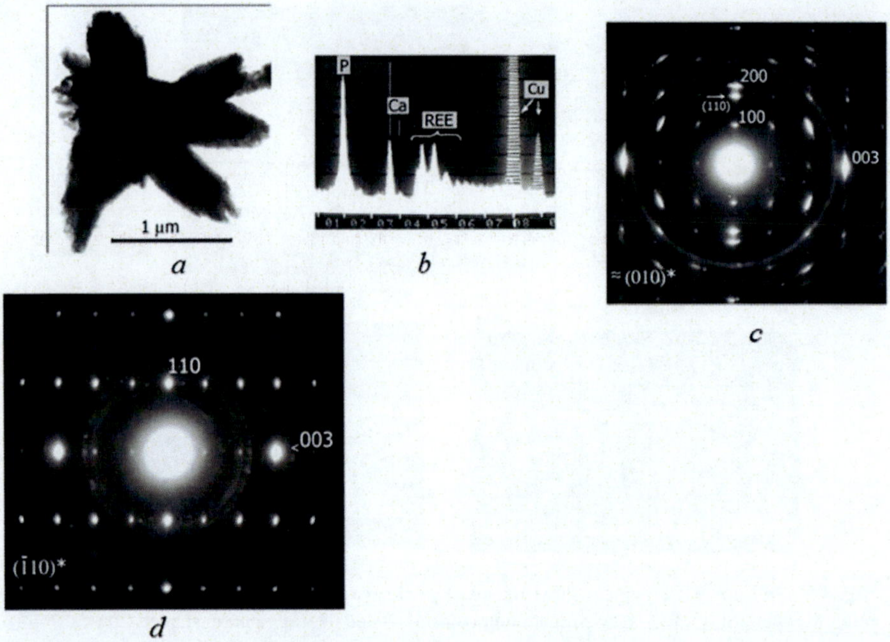

Fig. 4.4 Ca—rhabdophane variety (from apatite weathering crust, South Mongolia, sample of M. D. Dorfman). Microcrystal intergrowth image, TEM (**a**); ED- spectrum of composition (**b**); point SAED—patterns (**c**, **d**) Cu peaks from preparation

Calcium rhabdophane—variety of rare-earth phosphate with rhabdophane type structure—was found when studying apatite inclusions of Aldan shield (Fig. 4.3). The first studies (Guliy et al. 1990) were carried out on material that had undergone preliminary chemical treatment—dissolution of apatite in weak hydrochloric acid. Subsequent works on Aldan apatite study proved that performed specimen preparation does not lead to formation new mineral phases. Thus, natural origin of rhabdophane-like phosphate inclusions, mainly of calcium composition, is shown.

Else one rhabdophane variety containing significant calcium amounts in composition was noted in (Dorfman et al. 1993), here mineral composition $(Na_{0.02}Sr_{0.05}Ca_{0.35}Ce_{0.34}La_{0.29})_{1.05}\cdot(P_{0.94}S_{0.06})_1(O_{3.93}F_{0.07})_4\cdot 1.55H_2O$ was established (Fig. 4.4). Mineral elements ratio Ca: REE = 1: 2; Ca: P ~ 1: 3. The authors consider Ca′s entry into rhabdophane structure as result of heterovalent substitution according to scheme $2Ce^{3+} \rightarrow Ca^{2+}Ce^{4+}$, based on atomic quantities proximity of cerium and calcium in obtained formula. Cerium transition to tetravalent state is explained by its possible oxidation in conditions of discovery location (area crust of apatite weathering). However, in this case, it is not understood the simultaneously maintaining the unchanged lanthanum valence within the same structure. The isomorphism assumed by those authors also caused doubts in (Semenov 2000): "...isomorphism is unlikely, as Ce-content in mineral is normal". That is, cerium amount is not halved (compared with its usual contents in Rhabdophane), what inevitably would follow from this scheme. The authors proposed to consider this find as new rhabdophane variety, which idealized formula is $(Ca_{0.5}\ Ce^{4+}{}_{0.5})(PO_4)\cdot H_2O$. Such interpretation of obtained mineral analysis seems to be insufficiently substantiated. However, here it is important the very fact of rhabdophane detection, where divalent cations number exceeds previously known values by several times.

According to our results on Ca-rhabdophane study (Doinikova et al. 1993), Ca content is much higher and comparable with phosphorus content, which follows from composition spectra comparison (Figs. 4.3 and 4.4). Repeated finds of Ca-rhabdophane allow us to speak confidently about existence new Ca-variety in rhabdophane-like phosphates mineral group (https://www.mindat.org/min-53467.html).

Our study of Smirnovskite and Brockite museum samples, as well as new Ca-Rhabdophane and numerous samples of Ningyoite indicates that all mineral species that compose this group of aqueous phosphates are characterized by hexagonal symmetry. Pseudo-hexagonal (rhombic) symmetry that has been accepted here earlier in some species is absent. As result of conducted research, it was established that Rhabdophane mineral group, in addition to its own Ce-, Nd-, La-, and Y-varieties, includes 3 more minerals: thorium phosphate—Brockite identical to Smirnovskite, uranium phosphate—Ningyoite and unnamed rhabdophane-like calcium- REE phosphate. Data on minerals of Rhabdophane group tabulated (Table 4.1).

4.2.2 *Isomorphism in Rhabdophane Group Phosphates*

This section deals with the variability of the cationic composition in natural rhabdophane-like phosphates on the example of a typical representative of the group, studied in detail. Namely, the manifestation of cationic isomorphism will be considered in ningyoite.

As for isomorphism in the anionic part of the composition, the author does not exclude the possibility of replacing tetrahedral groups $[PO_4]$ in the structure of phosphates with groups [SiO4], $[SO_4]$, etc., which is allowed in structural mineralogy (Belov 1976). The recorded presence of S and, less often, Si in ningyoite composition indicates the possibility of isomorphism $S \leftrightarrow P$ and $Si \leftrightarrow P$ in tetrahedral PO_4 groups of anionic part of composition. The significant presence of sulfur is recorded in brockite and in Siberian ningyoite (Chap. 5).

According to data (Scharmova and Scharm 1994), isomorphic replacement of PO_4 with SO_4 and CO_3 is also permitted. Experimental evidence of anionic isomorphism in ningyoites composition is not yet sufficient for more detailed consideration. Partly for this reason, and mainly given the significant amount of water included in Rhabdophane structure, to compensate for the charge arising in the structure at isomorphic substitution of various elements, the author considers it preferable to operate by groups of OH to achieve valences balance.

Table 4.1 Mineral group of Rhabdophane (crystallochemical characteristics)

No	Name	Formula	Syngony, Sp. gr.	Cell parameters, Å	Data sources
1–3	**Rhabdophane**[a] -Ce, -La, -Nd	(Ce, La, Nd) $PO_4 \cdot H_2O$	Hexagonal, $P6_222$	$a = 6.98$ c = 6.39	Muto et al. (1959), Fleischer (1990)
	Ningyoite[a]	$Ca_{1-x}U_{1-x}REE_{2x}(PO_4)_2 \cdot 1{\div}2H_2O$ x = 0,1–0,2	Orthorhombic, P222	$a = 6.78$ b = 12.10 c = 6.38	Muto et al. (1959)
4	Ningyoite[b]	U_x (Ca, Fe)$_{2-x}$[P(O,OH)$_4$]$_2 \cdot$ nH_2O n = 1÷2 $Ca_{2-x}U_x$ [P(O, OH)$_4$]$_2 \cdot$ nH_2O	Hexagonal, $P6_222$	$a = 6.86$ c = 6.38	Belova et al. (1985, 1987); Doinikova et al. (1993)
	«Tristramite»	(Ca,U^{4+},Fe^{3+}) [(PO_4)(SO_4)(CO_3)] 1.5–2H_2O [Ca > > U, Fe; PO_4 > > SO_4, CO_3]	Hexagonal, $P6_222$	$a = 6.913$ c = 6.422	Atkin et al. (1983)
	Brockite[a] Brockite[b]	(Ca,Th) $PO_4 \cdot H_2O$ Th ≈ Ca; REE at background level $Ca_{2-x}Th_x$ [P(O, OH)$_4$]$_2 \cdot$ nH_2O	Hexagonal, $P6_222$	$a = 6.98$ c = 6.40 $a = 7.03$ c = 6.40 —	Fleischer (1962); Doinikova et al. (1993); Scharmova and Scharm (1994)
5	Grayite	Th-phosphate «some Ca, Pb; little U and Th»	Orthorhombic, Pseudohexagonal	–	Bowie (1957), Dooley and Hathaway (1961), Fleischer (1962)
	Smirnovskite	[Th_3 (Ce, Fe) (U, Mn, Pb)$_{0,1}Ca_1$]$\cdot$ [F_3OH_2]$_5$. [PO_4]$_4$. $5H_2O$	Tetragonal-?	–	Grigoriev and Dolomanova (1957); Dolomanova and Borisovsky (1979)
	Smirnovskite[b]	(Th, Ca) $PO_4 \cdot nH_2O$	Hexagonal, $P6_222$	$a = 7.12$ c = 6.47	Doinikova et al. (1993)
6	Ca-variety[b]	(Ca, REE) $PO_4 \cdot nH_2O$ Ca> > TR (La, Ce)	Hexagonal, $P6_222$	$a = 6.90$ c = 6.39	Doinikova et al. (1993), Guliy et al. (1990)

[a]Individual mineral species composed the group
[b]Minerals have been studied in this work

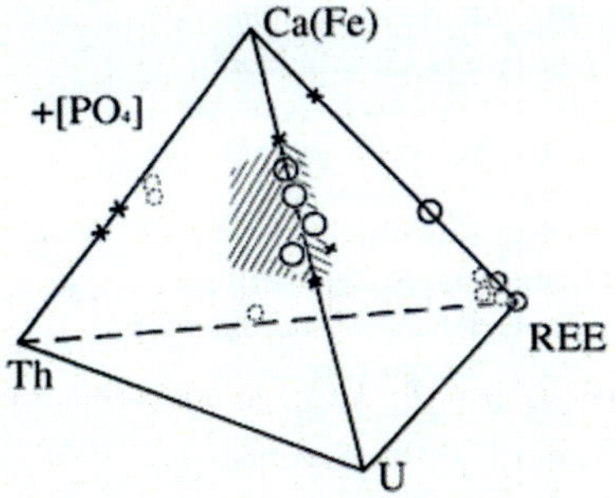

Fig. 4.5 Scheme-tetrahedron, showing the region of change in cation composition in phosphates with rhabdophane-like structure (Doinikova et al. 1993). Author's (crosses) and literary (circles) data; shading indicates region of compositions most characteristic for studied ningyoites

Detailed consideration of cation isomorphism of U-Ca-Th-REE in natural ningyoites allowed us to propose scheme characterizing the range of substitutions. Here, the theoretically possible schemes of isomorphic substitutions in the cationic part of rhabdophane-like minerals are considered by the example of ningyoite (Doinikova et al. 1993). For visualize consideration of cationic isomorphism, a tetrahedron scheme is proposed, denoting the range of cationic composition changes in the "phosphate" field (Fig. 4.5). The compositions of minerals, known from literary and our data are located within the proposed tetrahedron. It is existence area of rhabdophane-like aqueous phosphates with hexagonal elementary cell (same-type crystalline structures) in form of phosphate compositional tetrahedron.

In vertices of this tetrahedron elements entering into formula cationic part and, hence, occupying presumably identical structural positions in crystal lattice, are placed. Vertices correspond to the extreme members of isomorphic transformations within the structure—both real and hypothetical.

In tetrahedron tops there are cations REE, U, Th and Ca. Rhabdophane corresponds to REE top; ideally it is synthetic calcium-free Ce-, La-, Nd-, Y- analogues of rhabdophane. The Ca top indicates the direction of cationic composition changes leading to formation of independent mineral species: Ca-rhabdophane. Two other vertices conventionally designate the directions of cationic composition changes leading to formation independent mineral species: ningyoite (Ca-U composition) and brockite (Ca-Th composition). If rhabdophane structure is preserved, the hypothetical compositions corresponding to these vertices are obviously not achievable, what follows from the different valence of large cations REE^{3+} and U^{4+}, Th^{4+} which set the structure.

The variable Ca: U ratio in ningyoites composition indirectly indicates possibility of pure calcium phosphate existence in rhabdophane group. From reference data (Semenov 1963) almost all chemical analyses of natural rhabdophane have recorded the calcium presence. These facts led to allocation of independent top at compositional tetrahedron. The most convincing proof of expediency of such choices is discovery of rhabdophane-like predominantly Ca mineral (Guliy et al. 1990), as well as subsequent discovery of Ca-rhabdophane (Dorfman et al. 1993). Choice one of

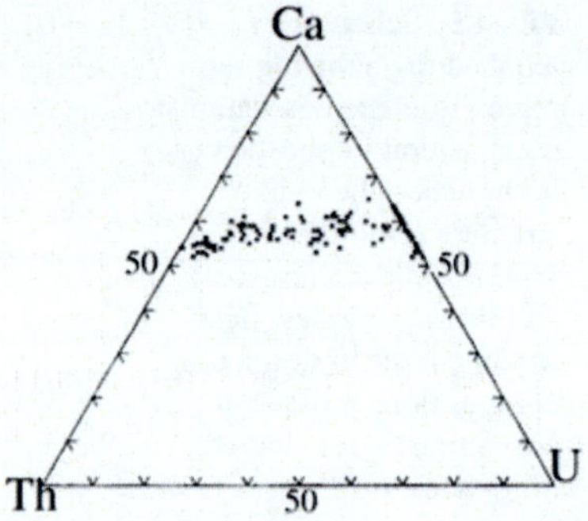

Fig. 4.6 Th-containing ningyoites (Strazh, North Bohemia, Czech Republic): phosphate compositions diagram demonstrates continuity of the compositions field between the end members of mineral series ningyoite-brockite and "Ca prevalence over U + Th" (Scharmova and Scharm 1994)

tetrahedron-diagram vertices for individual calcium mineral is justified by this. Iron (Fe^{2+}) also should be placed in the same peak because it established in ningyoite composition by ATEM data (Belov 1982) and quite often found in composition of other group members, sometimes in significant amounts (Kucha 1979).

The existence of brockite-ningyoite mineral series with isomorphism Th U↔ in cationic part of hexagonal phosphates composition is proved by our finding of Th-variety of ningyoite corresponding to intermediate member of this series. Numerous chemical analyses by Czech mineralogists of more recent finds (Fig. 4.6) have convincingly proved existence of continuous mineral series of brockite-ningyoite (Scharmova and Scharm 1994).

The monographs of E.I. Semenov (1963, 2000) report the discovery of minerals from series Rhabdophane-Smirnovskite-Brockite with almost equal content of REE and Th (so-called "Th-rhabdophane"), where REE composition can be either cerium or essentially yttrium (up to 50% of ΣREE). Together with the experimental results, these data allow us to assume the isomorphism Th ↔ REE.

ATEM research has shown that ningyoite almost always contains significant Ca amount. The ratio Ca: U in ningyoites composition, approaching the ideal 1:1, in different samples varies, as a rule maintaining Ca predominance. In extreme case, Ca: U ratio is close to values 3: 1. Literary data analysis shows that Ca is also always present in the other two phosphates in this group—Brockite and Rhabdophane. In samples studied by author, rhabdophane containing about 10% uranium and a little (≈ n%) calcium was found.

Compositions of rhabdophanes lie near REE top. Ningyoites occupy an interval in the middle of Ca-U rib/edge, and brockite (smirnovskite)—near the middle of Ca-Th rib. Diagram-tetrahedron clearly shows brockite-ningyoite mineral series (strip along U-Th edge, reflecting constant Ca presence). The composition of Ningyoite- Brockite mineral series is located in central part of flat with U, Th, and Ca vertices, in form of continuous series along U-Th edge (Fig. 4.6), indicating approximately equal proportions of 4-valent (U^{4+}, Th^{4+}) and 2-valent (Ca^{2+}) elements occupying the positions of 3-valent Ce (REE^{3+}).

Thus, compositions of all minerals of Rhabdophane group are located within proposed tetrahedron, indicating area of cationic composition change in "phosphate" field. In other words, the scheme presents peculiarities of cationic isomorphism in phosphate structures having common rhabdophane-like structural motif. Analysis of experimental and literary data, performed with such diagram help allows drawing the following conclusions:

- Rhabdofan's group does not have the continuity of substitution according to scheme Ce ↔ (U, Th) + Ca. Known isostructural minerals (rhabdophane and ningyoite/brockite) are extreme members of possible cations substitution according this scheme. Here there are no compositions with partial, incomplete isomorphism in single structural position. This indicates the discreteness, namely, variability divariantal filling this position in rhabdophane structure, which is manifested in absence of isomorphic series rhabdophanc-ningyoite (brockite) in nature.
- The presence of rare-earth elements (REE $< 10\%$) in some compositions of ningyoite- brockite series corresponds to position of these minerals on diagram inside the tetrahedron, near the plane U-Th-Ca. This fact allows us to assume the presence of structural positions, which are not involved in substitution by specified scheme. Presumably, these are individual Ca cationic positions that can be occupied by REE. The presence of such an individual Ca position can be explained by the author's find of natural rhabdophane, with small ($< 10\%$) content U and Ca.

The final solution to this question remains the structural analysis of natural mineral species of rhabdophane group, which could have clarified the synthetic rhabdophane structure previously defined (Mooney 1950).

4.2.3 Cationic Isomorphism Schemes

Here will be analyzed schemes of cationic isomorphism, theoretically possible in rhabdophane structure of ningyoite, i.e. possible substitutions, which are carried out during Rhabdophane group minerals formation. Based on review of well-known structure of rhabdophane $CePO_4{\cdot}H_2O$ as basic (initial), possible schemes of isomorphic (cationic) substitutions in Ca-U-phosphate structure are offered. The supposed mechanisms of heterovalent cationic isomorphism are based on experimentally fixed qualitative elemental composition of natural ningyoites. The probability of isomorphic schemes is based on objective data on mineral composition. Consider further the validity of each of mechanisms that based on elemental composition.

The most probable scheme of cationic isomorphism in rhabdophane-like ningyoite structure $2REE^{3+} \leftrightarrow U^{4+} + Ca^{2+}$ require preservation of quantitative equality of two- and four-valent elements.

In natural ningyoites composition, however, this equality is maintained only approximately, deviating in both directions. The ratio Ca: U, approaching ideal

1:1, in different samples varies, as a rule, maintaining Ca prevalence. (In extreme case ratio Ca: U is fixed close to 3:1) This scheme of cations interrelation most likely reflects general tendency in natural isomorphism processes, but does not explain them completely. This isomorphism mechanism does not explain, in particular, existence of ningyoite varieties with predominance Ca over U in composition of two or three times, as well as detected predominantly calcium rhabdophane variety.

The fluctuations in the composition changes observed in ningyoites, and some excess of Ca over U as well as the existence of Ca-rhabdophane author explains by the following type of valence compensation. Replacement of REE^{3+} by Ca^{2+} is compensated according to scheme REE^{3+} $[PO_4]$ $\leftrightarrow$ Ca^{2+} $[PO_3OH]$, i.e. by replacement of phosphate group oxygen by OH-group. In this case, no tetravalent elements are required to compensate for the charge. The significant predominance of ningyoites, in which Ca $>$ U indicates, that such mechanism works in nature along with above mentioned one. Such ningyoites formula (in absence REE): $Ca_{2-x}U_x[P(O,OH)_4)]_2 \cdot nH_2O$, where $x \leq 1$ (Belova et al. 1985). Same variant of isomorphism explains constant Ca presence in natural rhabdophane analyses. Representative results of quantitative EDS analysis of Ningyoite- Brockite series minerals (Scharmova and Scharm 1994) are in good agreement with such isomorphic substitution scheme (Fig. 4.6).

A slight predominance U $>$ Ca was recorded much less frequently. In this case, valence compensation for heterovalent replacement TR^{3+} by tetravalent atoms (U^{4+}, Th^{4+}) can be explained by replacement of structural water H_2O by hydroxyl group of oxygen, or by presence of vacant Ce positions in structure. However, this variant of isomorphism is less probable or has very limited effect, because in practice of author's studies there was only slight excess U over Ca in ningyoites composition.

During Rhabdophane group minerals formation, in nature most probable is combined effect of various isomorphic mechanisms. In order to clarify real mechanism of heterovalent atoms isomorphic replacement in rhabdophane-like structures, further research is needed, using modern physical methods, which will allow clarifying positions of different cations and hydroxyl in crystal lattice of minerals.

4.2.3.1 Relationship Between Isomorphism and Mineral Formation

The participation of hydroxyl OH-groups in valence compensation essentially means wide participation of hydrogen ions H^+ in this process. High probability of such compensation mechanism has real genetic rationale.

As it follows from publications majority, rhabdophane group phosphates, including ningyoite, are in nature in close association with organics that provides active participation of atomic hydrogen in mineral formation processes. Using ningyoite example, it is shown how reliability of proposed isomorphic substitutions schemes in mineral structure is verified by practice of studying this mineral composition.

In such consideration, variability of elemental composition of mineral (ningyoite) acts as criterion of the nature of isomorphic substitutions. The set of crystallochemical data obtained as result of various Ningyoites study serves to verify proposed hypotheses and schemes of structural isomorphic transformations. Another criterion is conditions of mineral's being in nature, i.e. natural paragenic mineral associations of uranium phases (interlayer with pyrite, "impregnation" of carbon material, etc.).

In addition, elemental composition of minerals can indirectly indicate mineral formation conditions through the nature of isomorphic substitutions in structure: as, for example, active H^+ role in ningyoite formation. Variations in elemental composition of mineral, emphasizing specifics of isomorphism indicate probable scheme of its implementation. Good example is cationic isomorphism in rhabdophane-like structure of ningyoite considered above.

4.2.4 On the Structural Position of Isomorphic Cations

The structure of rare-earth aqueous cerium hexagonal syngony phosphate was determined at its synthetic analogue (Mooney 1950). According to N.V. Belov's description (Belov 1976), Rhabdophane structure is characterized by *"main (more massive) architectural details—rods, in which large nine-vertex polyhedral, around the rare-earth Ce, and orthotetrahedra [PO4] alternate ... These columns are similar to those from Y- eight-vertex and P-tetrahedrons in Xenotime ... In Rhabdophanite [Rhabdophane] the columns are located less densely by sixes around hexagonal axis, and in created channels, ... along the channel axis, chain of H_2O molecules (screwed, twisted) stretches"* (Fig. 4.7).

Close eight-coordination is enough typical for Ca, Th, and U atoms in crystal structures. For example, in structures of Ca-garnet, gypsum, and anhydrite, Ca polyhedron is twisted Thomson cube. Th has similar coordination (by atoms O) in structure of Ekanite Ca-Th-silicate (Belov 1976). Ca-polyhedrons present in this

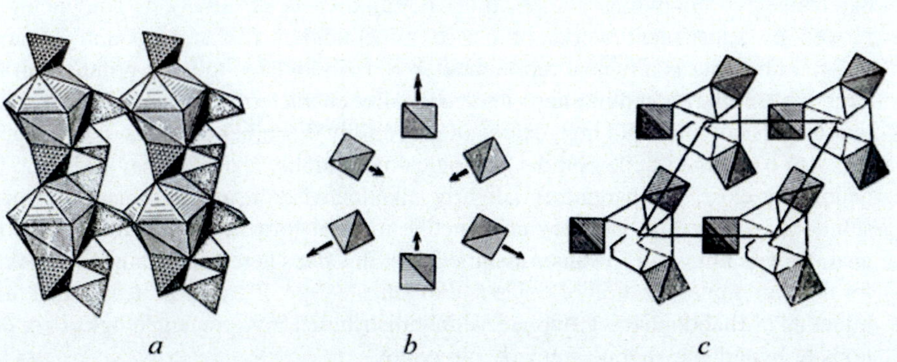

Fig. 4.7 Rhabdophane structure according to N.V. Belov (1976): (**a**) the main architectural rods from large 9- vertexes around Ce, interchanging with P-tetrahedra; (**b**) trigonal motive of the basic rods location in the structure of rhabdophane; (**c**) the main motive from hexagonal channels inlaid with P-tetrahedra (chains of H_2O molecules locate along the channel axis)

building are distorted (not strictly tetragonal) twisted cubes, where "the covers of the octagon are not exactly square and, besides, are broken diagonally".

Thorium and uranium in proper minerals are also characterized by eight-coordination of structural position: in oxides (thorianite, uraninite) polyhedron is cube, in silicates (thorite, coffinite)—an eight-vertex with trigonal faces. Thus, it is obvious that elements with very close to cerium (Ce) atomic radii dimensions—REE, U, Ca and Th—occupy common structural position in rhabdophane-like structures—in center of Ce-polyhedrons.

The placement of Ca in rhabdophane structure deserves closer consideration. Large eight-vertex around Ca^{2+} and REE^{3+} are also found in structure of Kainosite Ca-REE-silicate, but "*Ca and REE atoms here are not statistically mixed, but alternate, taking discrete independent positions ...* "(Belov 1976). Discrete columns create three-dimensional frame by linking in chess pattern forming channels along the *b*-axis.

In the structure of abovementioned Ekanite, despite the same eight-coordination and only small difference of coordination polyhedrons, Ca and Th are placed in separate structural positions, i.e. Ca does not occupy Th positions in the structure.

It can be assumed that in discussed group of phosphates rhabdophane-structured, similar to Kainosite and Ekanite, there are independent structural positions of Ca, in addition to it setting in Ce-polyhedron. Indirect indication of its position is continued presence of Ca in all minerals of Rhabdophane group. Only structural analysis of minerals of given group can confirm or dis-prove this assumption.

4.2.4.1 On Issue of Formation Conditions of Rhabdophane Group Minerals

The supposed possibility of continuous cation composition change in ningyoite-brockite minerals series based on uniformity of crystalline structures and wide range of changes in ningyoites compositions (Doinikova et al. 1993), is convincingly proved by subsequent works of Czech mineralogists (Scharmova and Scharm 1994). Ningyoite is a typical representative of rhabdophane mineral group, both in structural similarity and in micro-dimensionality characteristic of rhabdophane-like phases. It is possible with high degree of probability that other minerals of this group can also be formed under similar geological conditions. Namely, they can form at neutral or close pH parameter (slightly alkaline/weak acid) of mineral-forming solutions, near redox boundary at reductive mineral-formation medium. Based on analogy with ningyoite, it can be assumed that Eh values here will be higher than that for uranium mineralization of oxide and/or silicate type. It can also be assumed that minerals of rhabdophane group are widely distributed in various geological conditions of hypergenesis zone similarly ningyoite.

4.2.5 Summary

Consideration of Rhabdophane group minerals in the light of new crystallochemical data obtained by ATEM methods in combination with analysis of known reference data led to clarification of existing systematics in this mineral group. The following conclusions are interesting for mineralogy.

1. This group consists of seven mineral species, it is: *four* types of Rhabdophane (-Ce,—La, -Nd, -Y)—rare-earth aqueous phosphates which differ in predominance of one from listed components; Ca-U-phosphate—Ningyoite; Ca-Th phosphate—brockite; and unnamed rhabdophane-like Ca- mineral phase. Minerals, previously classified as tristramite and grayite are poorly studied analogues of ningyoite and brockite, respectively. Involvement reference museum samples to clarify existing Systematics in Rhabdophane mineral group showed priority of Russian mineralogists (Grigoriev and Dolomanova 1957) in discovery of natural thorium phosphate named Smirnovskite; it is analogue of later discovered (Fisher and Meyrowitz 1962), but studied in more detail Brockite, approved by CNMNC IMA as new mineral.
2. In composition of this group minerals, in formula cation part, elements REE, Ca, Fe^{2+}, U, Th are presented, which demonstrate possible isomorphism. List of isomorphic cations named in reference literature (Fleischer 1990) is supplemented by uranium U^{4+}. Noted earlier, Pb presence in this list is doubtful. According to formation conditions and crystallochemical data, it should be indicated ferrous iron form Fe^{2+} instead of the previously given ferric iron form Fe^{3+}.
3. Diapason diagram of cationic isomorphism within the group is presented as compositional tetrahedron with corresponding vertices (REE-Ca-U-Th). This diagram characterizes the manifestations of natural isomorphism within the limits of cerium structural position, which is corporate for cations. Scheme-tetrahedron shows the absence of isomorphic Rhabdophane—Ningyoite/Brockite series and demonstrates discreteness in replacement of Ce with (U/Th + Ca). This allows us to speak about divariant filling of Ce-polyhedrons structural position in rhabdophane structure that corresponding to the extreme variants of such substitution. Findings of minerals whose compositions slightly deviate from this pattern ($<10\%$) are considered as indication on possibility of individual Ca position existence in the structure.
4. The revealed structural and chemical characteristics of minerals and the natural isomorphism of their composition allow us to discuss conditions of mineral substances crystallization; these data indirectly indicate a probable scheme (mechanism) of isomorphic substitution.

4.3 On Question of Mineralogy Tetravalent Uranium U^{4+}

Speaking about uranium mineralogy, it should be emphasized that the behavior of this element in geological processes is fundamentally different for its different valence states. Minerals of tetravalent uranium U^{4+} are formed in a reductive geochemical medium and minerals U^{6+}—in an oxidative one. The chemical properties of U^{4+} and U^{6+} ions are so different that each of them forms compounds of a fundamentally different type, both in terms of their crystalline structure, morphology, and physico-chemical properties. Thus, it is necessary to consider independently, separately, the mineralogy of tetravalent and hexavalent uranium. In this monograph, we are talking about the mineralogy of the reduction zone of hypergenesis, and therefore about the minerals of the reduced form of uranium U^{4+}.

In the formation of minerals U^{4+} behaves like a typical cation, showing isomorphism with Th and *REE*, perhaps less frequently with Ca. In contrast, U^{6+} cation is individual (atypical) and is usually represented in nature as a linear ion $(U^{6+}O_2)^{2+}$—uranyl. The unique ability of U^{6+} to create a specific uranyl-ion determines the individuality of hexavalent uranium compounds.

A short excursus should be made here, considering the existence of pentavalent uranium in nature. For many years it was believed that U^{5+} could not create stable compounds, and therefore there are no minerals of pentavalent uranium in nature. In recent years, there have been publications about the existence of minerals with U^{5+} in the composition, but the question of the unambiguity of such uranium valence state determination remains open. The determination of the pentavalent form of uranium oxides (Teterin et al. 1981) remains questionable due to the heterogeneity and weak previous study of the reference samples. Data on the synthesis of stable compound of pentavalent uranium U^{5+}, and on the findings of natural mineral Wyartite with U^{5+} in the composition (Burns and Finch 1999) have been published. The conclusion about the presence of pentavalent uranium in the composition of this aqueous calcium-uranium uranyl-carbonate is based on the geometry of polyhedrons and the sum of valence bonds of one of the positions of uranium in the structure that requires the presence of U^{5+} for the electron-neutral balance of valences. This crystal-chemical proof is obtained by calculation, so it is indirect. Thus, in natural conditions only compounds of tetravalent and hexavalent uranium are known to date.

In uranium U^{4+} mineralogy the number of mineral species is very limited. There are more than 200 species of all uranium minerals; from them less than 5% are U^{4+}-minerals. There are: uraninite UO_2, brannerite $U(TiO_3)_2$, coffinite $USiO_4$, phospho-silicate (coffinite-like) $(U,Ca)[(Si,P)O_4]$, ningyoite $Ca,U(PO_4)_2{\cdot}nH_2O$. Less known are recently discovered by Czech mineralogists group (Plášil et al. 2011, 2013a, b) new uranous minerals from Jáchymov, Czech Republic,—sulphate behounekite $U(SO_4)_2{\cdot}4(H_2O)$ (2011) and arsenates vysokyite $U[AsO_2(OH)_2]_4{\cdot}4H_2O$ (2013), štěpite $U(AsO_3OH)_2{\cdot}4H_2O$ (2013).

Uranomolybdates—sedovite UMo_2O_8 and mourite UMo_6O_{20}—are complex oxides of U^{4+} and molybdenum with different ratios of uranium and molybdenum in their composition. In nature, these minerals are represented by highly dispersed

Table 4.2 Minerals of tetravalent uranium U^{4+}

Oxide	Uraninite	UO_2
Titanate	Brannerite	$U(TiO_3)_2$
Silicate	Coffinite	$U(SiO_4)$
Phosphates:	Ningyoite	$CaU(PO_4)_2 \cdot nH_2O$
Lermontovite group	Lermontovite	$U(PO_4)(OH) \cdot nH_2O$
	Vyacheslavite	— " —
	Urphoite	— " —
Phosphosilicate	–	(U,Ca) $[(Si,P)O_4]$
Sulphate	Behounekite	$U(SO_4)_2\ (H_2O)_4$
Arsenates:	Vysokyite	$U[AsO_2(OH)_2]_4 \cdot 4H_2O$
	Štěpite	$U(AsO_3OH)_2 \cdot 4H_2O$
Uranomolibdates	Sedovite	$U(MoO_4)_2$
	Mourite	$UMo_5O_{12}(OH)_{10}$

formations. These minerals are unstable, easily oxidized; as a result, U^{6+} and water molecules appear in sedovite (Sidorenko 1978).

The sedovite parameters determined by the debaegramm were confirmed by the SAED method (Å): $a = 6.42$, $b = 11.08$, c = 3.36; Z = 2; $\rho = 4.17$ ($d_{exp} \approx 4.2$). The X-ray powdergramm of mourite leaves no doubt about its mineral and structural identity; the indexing of the debaegramm was carried out with help SAED. Orthogonal cell parameters (Å): $a = 9.82$, $b = 7.17$, $c = 11.82$; Z = 2; $\rho = 4.17$ ($d_{exp} \approx 4.2$). According to G. A. Sidorenko (1978), these minerals are rarely found in nature because of their instability.

The lists of uranous minerals are given in the table (Table 4.2). In practice, only first five of it is known as the most important components of mineral raw materials for the nuclear industry. Therefore, special attention has been attracted to U^{4+} mineralogy and to deposits where these minerals are extracted on an industrial scale (it is mainly sandstone-type deposits).

Information about U^{4+} minerals has been greatly enlarged due to long-term study of black uranium ores, difficult for traditional diagnostics, with the help of highly local AEM methods. In the papers (Doynikova and Sidirenko 2010; Doinikova et al. 2009) the current state of tetravalent uranium mineralogy, which has developed in recent decades due to the wide use of crystallochemical studies by AEM methods, is considered. Until recently, this section of mineralogy was limited to three mineral species, namely: oxide *uraninite* and its morphological varieties (nasturane, uranium pitchblende), silicate *coffinite*, and titanate *brunnerite*. At the end of the twentieth century, industrially significant mineral species were supplemented by Ca-U^{4+}-phosphate *ningyoite*. Lermontovite mineral group, which is currently conditionally separated, combines U^4 -phosphates Lermontovite, Vyacheslavite, and Urphoite. Earlier single findings of sedovite and mourite known as U^{4+} molybdates (Sidorenko 1978) still remain insufficiently studied and are currently only of scientific interest. The supposed simultaneous coexistence in their structures of different forms of uranium—reduced (U^{4+}) and oxidized (U^{6+}), as well as the presence of the most oxidized form of molybdenum Mo^{6+}, seems to be geochemically unreasonable. Given the high dispersion of these mineral formations, their additional study is necessary.

4.3.1 AEM Data on Widely-Known Tetravalent Uranium Minerals

Uraninite UO_{2+x}. Crystal-chemical study of uranium-oxide mineralization of black ores by ATEM methods revealed the widest diapason of uraninite crystallinity degree. Different degree of uraninite crystallization is revealed by SAED pictures that were already mentioned in the first chapter of mineralogical section of monograph. It is interesting for mineralogists to know new information about the fact that in natural samples there are quite often electrono-amorphous formations of uraninite. As noted above, in the composition of poorly crystallized disperse uraninite often present an admixture of silicon, and the Si amount increases with a decrease in the degree of crystallinity. In this case, inclusion of Si in the mineral composition is caused by high sorption properties of the dispersed substance; it should not be associated with the structure of uraninite.

In U^{4+} mineralogy, new data obtained with the help of ASEM methods include findings of uraninite (nasturane) mineralization in the form of a mineraloid, a vitreous became solid gel. Mineraloid (opal-type formations) in mineralogical practice include mineral formations, which are hardened, initially colloidal substance, which does not have a crystalline structure.

This form, which is little known to mineralogists, is a solid mineral gel of uranium-oxide composition, which prevails in the ores of the Khokhlovskoye infiltration deposit, South Trans-Uralian (Dymkov et al. 2003, Volkov et al. 2005; Khalezov 2009; Velichkin et al. 2009a, b). The figure (Fig. 4.8) shows images of vitreous uraninite fragments reminiscent of hardened gel, with a characteristic conchoidal chipping, the spectrum of composition and the result of the analysis. In the same ores uranium-oxide "gel" is a mineraloid, it is found in the form of phyto morphoses and impregnation in carbonaceous lignite fragments (Fig. 4.8f, g). According to the analysis results (Table 4.3), zircon (4–6%), calcium (2–3%), silicon (1–4%), and sometimes vanadium (~1%) were found in this vitreous gel-nasturane.

Such mineraloid form of uranium-containing substance was first discovered by Yu.M. Dymkov et al. (2003); few finds of such uranium-bearing glasses in Khokhlovskoye deposit ores are considered; possible conditions of their formation are discussed. The majority of vitreous U-oxides contain water by indirect signs; their composition is close to that of uranium dioxide. The vitreous character of the mineral gel indicates the absence of substance crystallization, i.e. its amorphous state, which predetermines the presence of various impurities. Strictly speaking, such a non-crystalline uranium oxide compound corresponds to uraninite only in chemical composition; therefore, such a mineral substance is only conditionally related to uraninite. Content of impurities in vitreous "uraninite" (Al, Si, Ca, Zr, Ti, Si, Fe, S, and P) is insignificant, but their presence qualitatively characterizes geochemical features of uranium ore formation conditions. It would be desirable to note that in ores of paleo-channel type deposits (Khokhlovskoye) where the role of organic factor in ore accumulation is great, the presence of P and Ca at the level of 2–3% is constantly fixed in the composition of vitreous U-oxides; admixtures of Si and Al

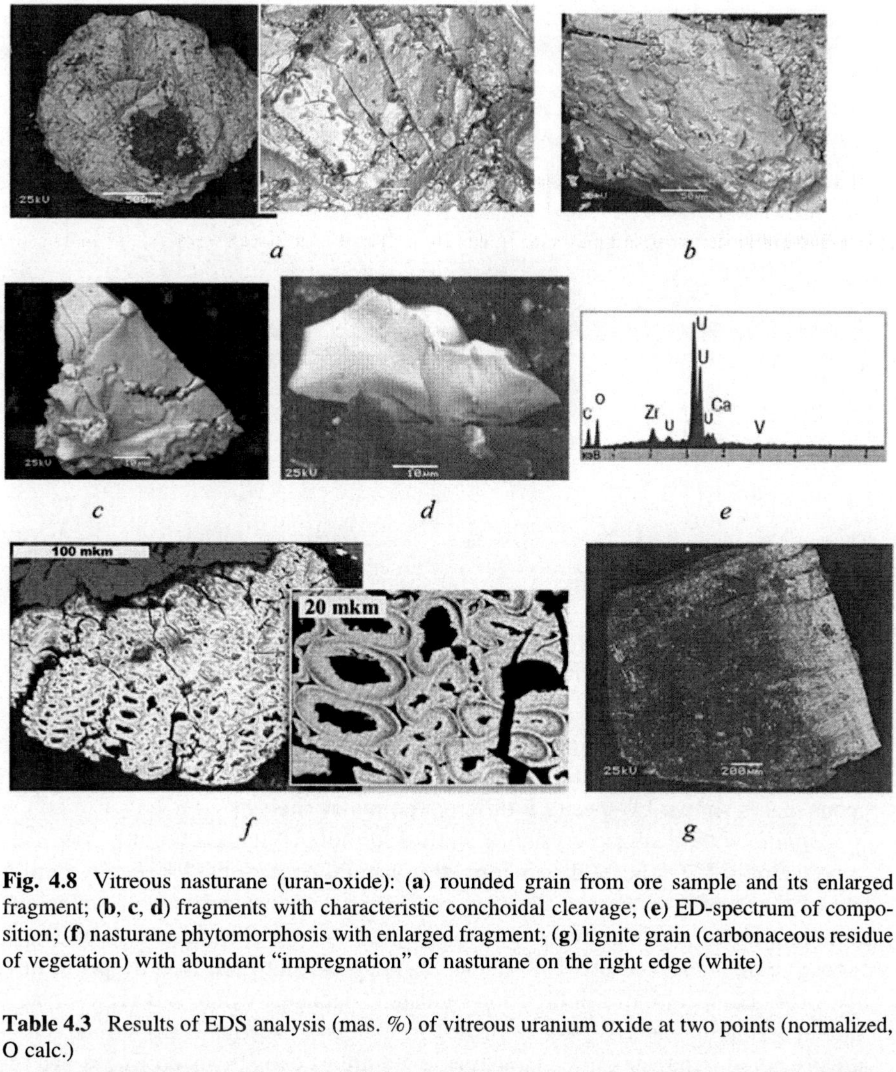

Fig. 4.8 Vitreous nasturane (uran-oxide): (**a**) rounded grain from ore sample and its enlarged fragment; (**b**, **c**, **d**) fragments with characteristic conchoidal cleavage; (**e**) ED-spectrum of composition; (**f**) nasturane phytomorphosis with enlarged fragment; (**g**) lignite grain (carbonaceous residue of vegetation) with abundant "impregnation" of nasturane on the right edge (white)

Table 4.3 Results of EDS analysis (mas. %) of vitreous uranium oxide at two points (normalized, O calc.)

	Element	Ca	V	Zr	U	O	Sum
1	Mas. %	2.41		4.64	79.66	13.30	100.00
2		2.45	0.44	6.28	76.96	13.87	100.00

in the composition of uranium-oxide mineraloid are not constant and do not retain such a stable ratio as phosphorus and calcium.

In exogenous ores of sandstone-hosted deposits uranium oxide is often associated with carbonaceous organic material, but much more often dispersed uraninite (nasturane) is adapt to cementation material, or acts as cement itself.

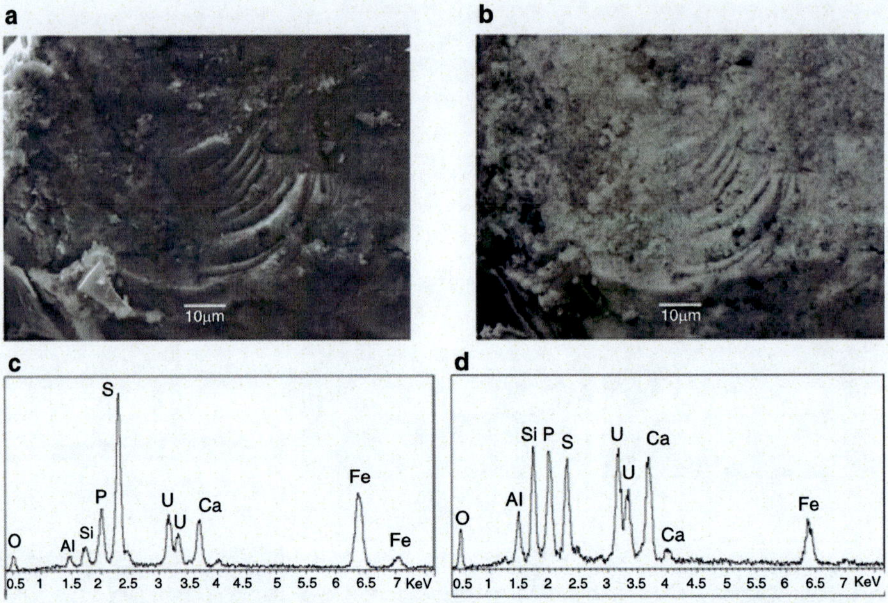

Fig. 4.9 Glassy mineraloid mass of ningyoite-pyrite composition: SEM- images in TOPO (**a**) and COMPO (**b**) modes; EDS spectra it different points (**c, d**)

Glassy mineraloid uranium-bearing mass have been recorded in one of the samples (Fig. 4.9). This vitreous mass of ningyoite-pyrite composition is heterogeneous that is shown by spectra it different points of analysis.

Coffinite with idealized crystallochemical formula $USiO_4$ is currently considered to be the only U^{4+} silicate that is formed in natural conditions. Practically until the end of the 1980s of the last century, coffinite was considered as an endogenous uranium mineral, and its presence in exogenous ores was considered a rare feature. Modern data show that this U^{4+}-silicate, as a significant part, is widespread in the black ores composition of infiltration sandstone deposits, where it is represented by micron formations and is closely related to the carbonaceous organic material. As already noted (Chapter 2), a characteristic feature of exogenous coffinites is a high degree of crystallinity, the ability to preserve the structure (i.e. point SAED patterns) even with a significant loss of uranium. In contrast to uraninite, the samples of exogenous uranium ores never contained electrono-amorphous coffinite.

Brunnerite U $(TiO_3)_2$ is the only titanate in which U^{4+} is a mineral-forming element. *Davidite*, often referred to by geologists as uranium titanate, is not actually uranium minerals. It is proved that from the crystallochemical point of view, it is a rare-earth *REE-titanate*, which conditionally belongs to the ilmenite group; in davidite composition uranium is an isomorphic admixture (Sidorenko 1978).

In mineralogical practice, for the study of structural characteristics of metamiktized brunnerite, preliminary ignition of the sample is traditionally used. As it was shown in the methodical section of the monograph (Chap. 1), the structural characteristics (parameters of the elementary cell) of brannerite in its natural unignited state were obtained due to the possibility of electron diffraction when study of X-ray amorphous sample. Uranium titanate brannerite was studied by author from Antey deposit, Streltsovsky ore field (Ivanova 1982). Based on the crystal chemical data of the ATEM, it has been established that the concept of "complex oxide" of U and Ti, which is widely used in the literature in the description of brannerite, is crystallochemical unreasonable. Unignited brunnerite is a titanate of tetravalent uranium. SAED microdiffraction data prove that non-calcined brannerite is tetravalent uranium titanate.

The family of tetravalent uranium phosphates, including a new mineral group of lermontovite, have been separated mainly on the basis of detailed crystallochemical studies by SAED method. Mineralogical Chapter 3 is devoted to the results of studying U^{4+}-phosphates. The minerals of this family are united by the general conditions of formation in the reduction zone of hypergenesis.

4.3.2 *Preconditions for Discovery of New Uranium Minerals*

The following is the actual material indicating the existence of previously unknown natural uranium compounds U^{4+}. These mineral phases, according to the existing norms, cannot yet be called new minerals, because their complete structural characteristics are missing. However, the quantitative analysis data obtained in ASEM does not allow doubting the individuality of their composition. The results obtained experimentally do not exclude the possibility of appearance of new uranyl-titanium minerals.

4.3.2.1 Mineral Phase of the Phosphosilicate Composition

The phosphosilicate phase of uranium U^{4+} was established in association with the disperse uranium oxide during study sandy ores of paleovalley (channel type) uranium deposit Dalmatovskoye by ASEM methods (Doinikova et al. 2009). This U^{4+} mineral phase is predominant in the studied ore intervals; it is usually confined to the changed feldspar. Microcrystals slightly elongated, lentil-shaped, a few microns in size, often form aggregates (Fig. 4.10), less often—collomorphic excretions. At this stage of research the idealized formula of the studied phase is presented as $(U,Ca)[(Si,P)O_4]_2$, where U:Ca $\approx$ 3:1 and Si:P $\approx$ 3:1. The ratio of cationic to anionic component (U,Ca): (Si,P) = 1:2, which is half as much as in coffinite, where U:Si = 1:1. As the relief rather than polished objects were studied and analyses results were normalized, it is required further study of polished samples composition; the question of OH-groups or water presence in this phase also remains open. Since the quantitative EDS analysis (Table 4.4) showed (in atomic units) a significant excess of silicon over phosphorus (three times), this phase can be characterized

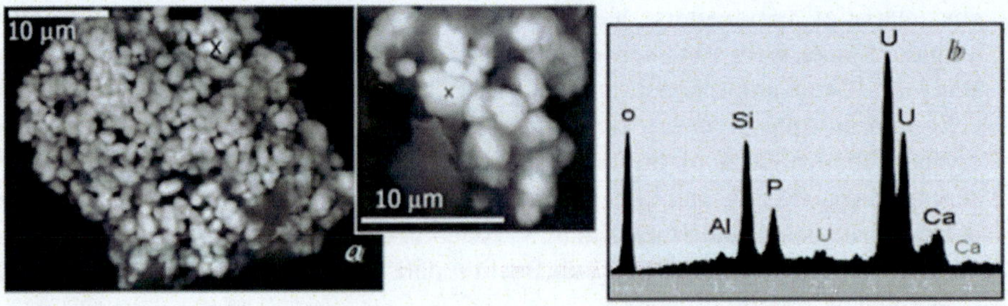

Fig. 4.10 Coffinite-like uranium phospho-silicate: (**a**) aggregates of microcrystals, (**b**) ED-spectrum of composition; (x)—points of analysis, spectra are the same

as a silicate of uranium containing phosphorus, or rather—as phosphosilicate of uranium (Doynikova et al. 2014, 2015).

The morphology of the microcrystals of the new uranium phase is close to the forms of hypergenic U^{4+}-minerals: coffinites and ningyoite. In contrast to ningyoite, this U-phosphosilicate contains a stable predominance of U over Ca. From P-containing coffinite the described phase differs, first of all, by the constant presence of calcium, as well as by the stable Si/P ratio. The revealed features of the composition allow speaking about individuality of this mineral phase.

A preliminary study of this phase by microdiffraction (SAED) has shown that diffraction patterns (as circular and so point) are characteristic of coffinite, which makes it possible to call the mineral coffinite-like. To determine the structural parameters, further diffraction studies with reflexes intensities analysis are required. The obtained data suggest that the development of the family of U^{4+}-phosphates due to the appearance of a branch of phosphosilicates is expected. Probably, to already known minerals of tetravalent uranium it will be possible to add one more—U^{4+} phosphosilicate.

4.3.2.2 About Uranium-Titanium Mineralization

The issue of uranium minerals of titanium is of interest to mineralogists because it still remains poorly understood. For a long time in the practice of mineralogical research, when the correlation between uranium and titanium in mineral formations was revealed, there was no doubt that there was a choice of only two minerals—brannerite and davidite. As noted above, only brannerite is titanate (davidite is REE oxide).

Relatively recently was been discovered a new mineral of uranium and titanium—*holfertite*, Ca-containing uranyl titanate with the idealized formula $(UO_2, Ca)_2\ TiO_4{\cdot}4H_2O$. Mineral was discovered by D.I. Belakovskiy and co-authors in the rhyolites of Utah, USA, in the form of non-luminescent tubular crystals (n·мм) of bright yellow color (Belakovskiy and Pautov 2002; Belakovskiy et al. 2006). Thus, at present in mineralogy there is one mineral of tetravalent uranium U^{4+} (brannerite)

Table 4.4 Results of EDS analysis of coffinite-like uranium phosphosilicate in different points

Elem.	1					2					3				
	Mas.%	σ	At. %	Form.	Mas.%	Mas.%	σ	At. %	Form	Mas.%	Mas.%	σ	At. %	Form.	Mas.%
Na	0.4	0.17	0.53	Na2O	0.54	–			Na2O		0.83	0.29	1.22	Na2O	1.11
Mg	–					0.17	0.15	0.23	MgO	0.29	–			MgO	
Al	4.83	0.15	6	Al2O3	9.13	5.6	0.16	6.82	Al2O3	10.59	2.89	0.21	3.63	Al2O3	5.47
Si	12.37	0.19	14.75	SiO2	26.46	12.52	0.19	14.64	SiO2	26.78	12.14	0.38	14.65	SiO2	25.98
P	3.34	0.14	3.61	P2O5	7.64	3.12	0.14	3.31	P2O5	7.14	3.61	0.2	3.95	P2O5	8.27
S	0.72	0.1	0.76	SO3	1.81	0.84	0.1	0.86	SO3	2.09	–			SO3	
K	0.46	0.19	0.4	K2O	0.56	–			K2O		–			K2O	
Ca	2.93	0.14	2.45	CaO	4.09	2.69	0.14	2.2	CaO	3.79	2.73	0.18	2.31	CaO	3.82
Fe	0.74	0.13	0.44	FeO	0.95	0.72	0.16	0.43	FeO	0.93	0.27	0.19	0.16	FeO	0.35
U	42.25	0.67	5.95	UO2	47.93	42.28	0.67	5.83	UO2	47.96	45.11	1.24	6.42	UO2	51.18
O	31.07	0.44	65.07		99.11	32.15	0.54	65.28			28.61	1.16	60.62		
Σ	**99.11**					**100.1**					**96.19**				

Fig. 4.11 Uranium-containing Ti-mineral (pseudorutile?). Nodular micro-spherulites of new formations (**a**); enlarged fragment (**b**); ED- spectrum and analysis result (normalized)

Table 4.5 Result of composition EDS analysis one of micro-spherulites (normalized)

Element	O	Al	Si	P	S	Ca	Ti	V	Fe	U	Sum
Mas. %	37.94	1.44	1.03	0.50	0.38	0.36	43.78	2.43	2.91	9.25	100.00
Atom. %	66.78	1.50	1.04	0.45	0.34	0.25	25.74	1.35	1.47	1.09	

and one mineral of hexavalent U^{6+} (holfertite). In view of the above, mineralogists should avoid hasty conclusions when dealing with U-Ti-mineralization.

Our experience of studying hydrogen uranium ores from infiltration sandstone-type deposits shows the frequent presence of Ti-mineralization, which sorbs uranium. U-Ti mineralization of hydrogen ores attracts attention and requires careful consideration, because not only creates geochemical specificity of ores, but also often has industrial value. For example, this specificity is characteristic of paleovalley type uranium deposits into Trans-Urals (Dalmatovskoye and, to lesser degree, Khokhlovskoye); so as deposits of a similar type from the ones of Bukinay group (Central Kyzyl-Kum uranium-ore province).

It is a fixing uranium mineral phase, which is a product of ilmenite change in hydrogen ores and is formed in the form of crusts on its "leached" surface. Very small kidney-shaped new formations of this uranium-containing titanium phase are rare (Fig. 4.11, Table 4.5). Its' high dispersity and close connection with ilmenite make it impossible to select monofraction for traditional methods of mineral diagnostics. Involvement of EM methods (both transmission and scanning) in the study of these ores material was primarily aimed at the study of uranium minerals themselves; the study of uranium-containing minerals received less attention. To date, the mineral form of this substance, contains uranium and titanium, has not been sufficiently studied.

It should be noted that the question of crystallochemistry of change products of natural Ilmenite in a similar geological setting, have been studied by Russian crystallographer and mineralogist G.K. Krivokoneva. Representative X-ray

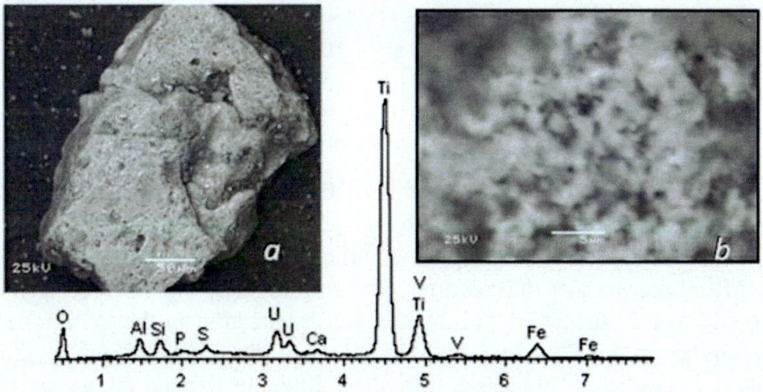

Fig. 4.12 Uranium-containing products of ilmenite change. Grain of altered ilmenite (**a**), enlarged fragment of its surface (**b**), ED- spectrum and result of composition analysis at the brightest point (normalized)

diffraction studies were carried out on samples of Ti-Zr sands of the Beshpagirskoye field (Krivokoneva 2005; Krivokoneva and Vasilyev 2006). It is established that the mineral phase forming a crust on titanate is a pseudorutile $Fe_2^{3+}Ti_3^{4+}O_9$. Such diagnostics seems to be more reasonable than the supposed attribution of ilmenite change products to anatase. Therefore, the *fixation of uranium in the altered ilmenite surface occurs in the transition conditions* $Fe^{2+} \rightarrow Fe^{3+}$.

When studying U-containing titanates of one of Bukinay deposits (Belova et al. 1991) by ATEM methods, it was noted that in a number of samples with uranium pitchblende, where titanium hydroxides are present, an extremely low ratio of U^{+4}/U^{+6} is recorded. It's linked to the *presence of hexavalent uranium in titanium hydroxides*. In this case, from geochemistry viewpoint it is logically, that the relationship between the most oxidized forms of iron and uranium, accompanying the transformation of ilmenite into pseudo-rutile.

The elemental composition study of U-containing Ti-minerals in Dalmatovskoye and Khokhlovskoye deposits by ASEM methods has shown that ilmenite leaching is accompanied by Fe removal and U accumulation in the resulting crust-mass, covering ilmenite. The composition of altered ilmenite, containing uranium, is strongly depleted in iron and approaches the formula $(Ti,Fe)O_2$ with ratio Ti:Fe = 2:1 (Doinikova et al. 2009). According to our data, the uranium content is proportional to degree of ilmenite destruction, which confirms the previously established (Krivokoneva 2005) inverse dependence of uranium and titanium content in the grains of altered titanium mineral. There is direct dependence of the relative uranium amount on Fe removal degree from ilmenite composition. This indicates *connection between Fe removal and U supply;* uranium is fixed by the "destroyed" surface of the titanium phase, presumably in sorption form.

The most altered areas on ilmenite surface can be considered as submicron uranium phase formations at the initial stage of mineral formation. The composition of this phase shows significant loss of iron and the presence of Si, P, Ca, Al admixtures in addition to uranium (Fig. 4.12, Table 4.6). The peaks ratio allows

Table 4.6 Result of EDS composition analysis in the brightest point (normalized)

Element	O	Al	Si	P	S	Ca	Ti	V	Fe	U	Sum
Mas. %	38.75	2.06	1.55	0.35	0.63	0.47	42.52	1.78	3.99	7.89	100.00
At. %	66.84	2.11	1.52	0.31	0.54	0.33	24.49	0.96	1.97	0.91	

only assuming the presence here of both uranium oxide and may be the above described phosphosilicate uranium phase.

U-Ti- phase in the form of plate-submicron formations was found during the study of Dalmatovo ores thin sections. The confinedness of this extremely small and rare phase to the surface of Ti-minerals and its location in the intergranular space leave doubts in the "purity" (individuality) of the composition spectrum. The oxidative character of processes that alters ilmenite surface allows us to assume here the initial stage of one of uranyl minerals (titanate or uranite) onto titanium oxide. If we assume that U-titanate is fixed here, then, unlike holfertite, there is no calcium in its composition (within the accuracy of measurement), and the ratio U/Ti shows a significant excess of titanium over uranium. The assumption of holfertite pioneers (Belakovskiy et al. 2006) about the possibility of its formation as a result of uranium oxidation from U^{+4}-minerals to hexavalent state of U^{+6}, would explain the low ratio U/Ti.

The above data allow only presumably speaking about the presence of uranyl titanate (possibly in the initial stage of formation) in the studied samples of Dalmatovo ores. This hypothesis requires additional confirmation. However, it is obvious that the solution to the question of the mineral form of uranium in the composition of the products of change of Ti-minerals requires more detailed study, rather than hasty assignment of the mineral to the brannerite. Solution this issue may lead to the discovery of new mineral species of uranium.

4.3.3 About Ca Place in Tetravalent Uranium Minerals

The question of the form of the presence of Ca in the U^{4+}-minerals has long attracted attention and remains a subject for discussion. It is noteworthy that both phosphates and coffinite-like phosphosilicate with a significant amount of phosphorus in the composition contain Ca equally with U^{4+}. Can we talk about isomorphic substitution here?

Isomorphic substitution of U^{4+} by Ca^{2+} with equal ion radii always caused doubts due to the crystallochemical individuality of everyone. The place of calcium in uranium minerals has been in dispute for a long time. Initially, the equality of ionic radii and isoctructurality of uraninite with fluorite was accepted as a justification for isomorphic replacement of U^{4+} and Ca^{2+}. There was also no doubt that U^{4+} substitutes Ca^{2+} in the structure of apatite. This version has been disproved by the results of technological processing of ores in which U^{4+} is associated with apatite.

When pure apatite was released, they didn't get the expected uranium concentration. Therefore, uranium is not a structural admixture of apatite.

However, it is known that uranium enters the pyrochlore structure with formation of uranium pyrochlore (Ca,Na,U)(Ti,Nb,Ta)O_6(F,OH), in which uranium U^{4+} statistically occupies the Ca^{2+} position in the structure. The consequences of this phenomenon are also traced by the X-ray method when uranium in composition increases: there is disordering structure of pyrochlore, up to destruction (possibly through metamict state), and stoichiometry disturbance. To maintain the balance of the charge in the structural motif of AB_2O_6 (OH,F) at the increase of uranium amount is the enrichment of the mineral with titanium Ti in position B (Ta,Nb). As a consequence, there is a disturbance of stoichiometry in the structural motif. In addition, the crystal building of the mineral itself is disrupted. Speak by the expansion of the X-ray diffraction lines, as the structure is enriched with uranium, first there is a decrease in the size of the unit cell (2 times), then three-dimensional order disturbance, then full metamictization (data by G. A. Sidorenko).

Probably, at ordered resettlement of U^{4+} and Ca^{2+} on structural positions, an individual crystalline structure will be established, that, probably, takes place in structure of ningyoite. The second example, presumably, will be the structure of a new phosphosilicate phase similar to coffinite (see above). To suppose that the resettlement of U^{4+} and Ca on structural positions of cations is ordered, allows sufficiently persistent, clearly fixed, ratio of uranium and calcium, both in the idealized composition of ningyoite and in the composition of phosphosilicate. For ningyoite (CaU)PO_4 the ratio U/Ca = 1/1, and for the coffinite-like phase (U,Ca) [(Si,P)O_4]$_2$, containing significant amounts of phosphorus, the ratio U/Ca = 3/1.

Coffinite-like Ca-U^{4+}-phosphosilicate is interesting not only for its mixed anionic composition, but also for its high content of Ca, which along with uranium becomes a mineral-forming element. The anionic isomorphism of SiO_4—PO_4 was experimentally shown earlier in the study of P-containing coffinites: Si and P (insignificant quantities) coexisted in a single structure providing characteristic electron diffraction patterns (Belova et al. 1980).

Let us consider a hypothetical change in the chemical composition from $USiO_4$ coffinites to ningyoite (Ca,U)PO_4. From the crystallochemical point of view, the complete replacement of Si with P is accompanied by the appearance of calcium in the cationic part. The crystalline structure changes cardinally from tetragonal to hexagonal. Therefore, here there is not an isomorphism of U^{4+}—Ca, but the formation of the cationic part of the composition of two elements. A certain ningyoite-specific predominance of Ca over U (a variable) can be considered as one of the variants of Ca location in the structure. In general the ningyoite composition overall regularity is retains U: Ca = 1: 1 (Doinikova 2003).

Let us then consider the change in the composition from $USiO_4$ coffinite to U^{4+}-phosphosilicate (U,Ca)[(Si,P)O_4]$_2$. In the cationic part of the latter, the ratio U: Ca = 3:1. Compared to ningyoite, the relative further excess of U over Ca is accompanied by the manifestation of isomorphism of Si–P in the anionic part of the composition. Thus, we can say that the combination of silicate SiO_4 and phosphate PO_4 groups in the structure of U^{4+}-mineral leads to the indispensable appearance of Ca ion here. In the composition of phosphosilicate the valence balance is achieved by an equivalent change in the composition, with the same ratio 3:1, by the replacement of U for Ca and Si on P.

Thus, when phosphosilicate mineral species appear in the U^{4+}-phosphate family, the anionic part of the composition is transformed, the cation component changes too, and U^{4+} necessarily "cooperates" with Ca in structure.

According to the data available to date, the individual predominantly structural position of Ca, rather than isomorphic, in tetravalent uranium minerals is more likely. This kind of "paired" replacement takes place in the crystallochemical group of apatite-britholite (apatite is typical phosphate, and britholite is its silicate structural analogue). It should be emphasized that it is the silicate analogue with constant phosphorus content that has U^{4+} in its composition (also rare-earth elements and Th).

What can be saying about the non-lattice Ca? Obviously, it is present as admixtures (~ 2% Ca) in the composition of non-structural harden uranium-oxide gels, the so-called uranium-bearing glasses. Calcium, which, according to data (Khalezov 2009), is fixed in the composition of uraninite simultaneously with phosphorus (up to 3.3% CaO, up to 2% P_2O_5), is logically can be associated not with uraninite structure, but with emergence of phase heterogeneity (formation a new micro phase).

4.4 Conclusions

Developing knowledge on uranium mineralogy at the expense of uranous U^{4+}-minerals in recent decades is due to the wide use of crystallochemical studies by ATEM.

- From 2000s industrially significant mineral species (uraninite, coffinite, and brunnerite) were supplemented by CaU-phosphate ningyoite.
- New Family of tetravalent uranium phosphates, including a new lermontovite mineral group (including also vyacheslavite and urphoite) have been separated mainly on the basis of crystallochemical studies by SAED method. Lermontovite mineral group (Lermontovite, Vyacheslavite, and Urphoite) have been currently selected conditionally. Subsequent finds of this group minerals will make it possible more detailed analysis their chemical composition and refinement structural parameters. Then each of them may become a founder of new mineral group. Obtained data allows us suggest to expansion of this uranium phosphates mineral Family due to branch (group) of phosphosilicates.
- On example of brannerite it was shown possibility study of X-ray amorphous samples; set of its crystallochemical data (composition and structural characteristics) have been obtained in its natural unignited state.
- Unknown earlier mineraloid form of nasturane (vitreous solid gel) has been found.

All above listed obtained crystallochemical data shows that ATEM with SAED microdiffraction is powerful instrument for uranous mineralogy investigation.

References

Atkin D, Bashatn IR, Bowles JW (1983) Tristramite, a new calcium uranium phosphate of the Rhabdophane group. Mineral Mag 47:393–396

Belakovskiy DI, Pautov LA (2002) A new hydrous uranium titanate from Thomas Range, Utah, USA. In: International mineralogical association. 18th general meeting. Edinburgh. p 135

Belakovskiy DI, Pautov LA, Sokolova E, Hawthorne FC (2006) Holfertite, a new hidroxil-hydrated uranium titanate from starvation canyon, Thomas range, Utah, USA. Mineral Rec 37:311–317

Belov NV (1976) Essays on structural mineralogy. Nedra, Moscow, p 344. [in Russian]

Belov NV (1982) Main stages of development of ideas in crystallography. In: Eliseev EN, Sachkov YV, Belov NV (eds.) Streams of ideas and regularities of natural science development. Nauka. pp. 159–172. [in Russian]

Belova LN, Gorshkov AI, Ivanova OA, Sivtsov AV (1980) Nature of so-called P-bearing coffinite. Dokl AN SSSR 255(2):428–430. [in Russian]

Belova LN, Gorshkov AI, Ivanova OA et al (1985) Ningyoite in the light of new experimental data. Proc USSR Izvestiya SSSR Ser Geol 3:101–109. [in Russian]

Belova LN, Gorshkov AI, Doynikova OA, Sivtsov AV (1987) On analogy between tristramite and ningyoite. Izv Akad Nauk SSSR Ser Geol 11:136–137. [in Russian]

Belova LN, Gorshkov AI, Sivtsov AV (1991) Sorption of uranium by titanium hydroxides as one of the sources for creating industrial concentrations of uranium in epigenetic deposits of sandstone-type uranium. Dokl AN SSSR 316(1):191–193. [in Russian]

Bowie SHU (1957) Summary of progress. Geol. Servey Gt. Brit. for 1956. London, p 67

Burns PC, Finch RJ (1999) Wiartite: crystallographic evidence for the first pentavalent uranium mineral. Am Miner 84(9):1456–1460

Doinikova OA (2003) Genetic crystal chemistry of mineral components in uranium blacks. Geochem Int 41(12):1214–1220

Doinikova OA, Gorshkov AI, Belova LN et al (1993) Questions on systematics of phosphates Rhabdophane group. Zapiski VMO 3:79–88. [In Russian]

Doinikova OA, Solodov IN, Chertok MB (2009) Mineral composition of uranium ore at the Dalmatovo deposit, Russia. Geol Ore Deposits 51(6):505–514. https://doi.org/10.1134/S1075701509060051

Dolomanova EI, Borisovsky SE (1979) New data on chemical composition of Smirnovskite. New data on minerals of the USSR. Miner. Museum AN USSR. Nauka. pp 168–171. [in Russian]

Dooley JR, Hathaway JC (1961) Two occurrences of thorium-bearing minerals with rhabdophane-like structure. US Geol. Survey Prof. Paper 424–C, pp 339–431

Dorfman MD, Gorshkov AI, Nechelyustov GN (1993) Calcium rhabdophane—a new variety. Doklady RAS 331(5):583–586. [in Russian]

Doynikova OA (2012) Uranium mineralogy in hypergenesis reduction zone (according to electron microscopy data). Fizmatlit, Moscow, p 216. [in Russian]

Doynikova OA, Sidorenko GA (2010) To mineralogy of tetravalent uranium. New data on minerals, vol 45. Min. Muz. RAS, Moscow, pp 76–90

Doynikova OA, Sidorenko GA, Sivtsov AV (2014) Phosphosilicates of tetravalent uranium. Doklady Earth Sci 456(2):755–758. https://doi.org/10.1134/S1028334X14060336

Doynikova OA, Sidorenko GA, Mokhov AV, Sivtsov AV (2015) New data on phosphosilicates of tetravalent uranium. New data on minerals, vol 50. Min. Muz. RAS, Moscow, pp 11–19. (in Russian, English abstract)

Dymkov YM, Volkov NI, Doynikova OA (2003) Genetic features of uranium oxides from the Khokhlovskoye Deposit (Trans-Urals). Urals Mineralogy. In: Proc. 2003. Mias: vol. II. p. 62–67 [in Russian]

Fisher FG, Meyrowitz R (1962) Brockite, a new calcium thorium phosphate from the Wet Mountains, Colorado. Am Miner 47(11/12):1346–1355

Fleischer M (1962) New mineral names. Am Miner 47(3–4):419–420

Fleischer M (1990) Glossary of mineral species. Mir, p 306 [in Russian]

Frank-Kamenetsky VA (1978) Structural typomorphism and modern mineralogy. Mineral Collect Lviv Univ Lviv 2(32):19–22. [in Russian]

Frank-Kamenetsky VA (1985) N.V. Belov-the founder of domestic Russian structural Mineralogy. In: Crystal chemistry and structural typomorphism of minerals. L-d: Nauka, pp 5–11. [in Russian]

Grigoriev IF, Dolomanova EI (1957) Smirnovskite–a new mineral from the group of aqueous fluorosilico-phosphates of thorium. ZVMO 5:607–621. [in Russian]

Guliy VN, Doynikova OA, Furuta T, Sivtsov AV (1990) Mineral inclusions in Apatite of metamorphic rocks and ores of the Aldan shield. Mineralogical digest of Lviv state Univ. Lviv. 2 (44), 69–73. [in Russian]

Hydrogenic Uranium Deposits (1980) In: Perelman AI (ed.) Atomizdat, Moscow, p 270. (in Russian)

Ivanova OA (1982) Electron microscopic study of fine-dispersed uranium minerals. PhD dissertation, IGEM RAS. Moscow. 172 p. [in Russian]

Ivanova OA, Belova LN, Sivtsov AV, Cherkasov AV (1985) Yttrium-containing ningyoite. VMO Conf."Basic problems of theoretical and applied mineralogy". Abstr, p 144. [in Russian]

Khalezov AB (2009) Uranium deposits in river paleovalleys of the Urals region. M.: VIMS. 145 p. (in Russian)

Krivokoneva GK (2005) X-ray quantitative phase analysis (XRQPhA) of black heavy concentrates from the ore Ti-ZrSands: the case of the beshpagir deposit. RITS VIMS, Moscow. [in Russian]

Krivokoneva GK, Vasilyev AT (2006) The use of XRQPhA for estimation of resources of ore minerals in TiZr-sands of the beshpagir deposit. In: Proceedings of conference on Ti-Zr deposits of Russia and perspectives of their development. IGEM, Moscow, pp. 29–32. [in Russian]

Kucha H (1979) $Fe^{2+}Th(PO_4)_2$ monoclinic, $Fe^{2+}Th(PO_4)_2 \cdot H_2O$ hexagonal, $Fe^{2+}{}_{1-x}Th_{1-x}(REE,Fe^{3+})_{2x} \cdot (PO_4)_2 \cdot 1 \div 3H_2O$ orthorhombic and $Fe^{2+}{}_3(H_2O)(PO_4)_2$ monoclinic – four new minerals from Poland. Miner Polonica 10:3–25

Mineralogical Tables (1981) Handbook. Nedra. 399 p. [in Russian]

Mooney RC (1950) L X-ray diffraction study of serous phosphate and related crystals. I. Hexagonal modification. Acta Cryst 3(Pt 5):337–340

Muto T, Meyrowits R, Pommer AM, Murano T (1959) Ningyoite – a new uranous phosphate mineral from Japan. Amer Mineral 44(5–6):633–639

Perelman AI (1968) Geochemistry of epigenetic processes (hypergenesis zone). Nedra, p 331. [in Russian]

Perelman AI (1989) Geochemistry. Higher School, Moscow, p 574

Perelman AI, Kasimov NS (1999) Landscape geochemistry. Astrea-2000, Moscow, pp 768. (in Russian)

Plášil J, Fejfarová K, Novák M, Dušek M, Škoda R, Hloušek J, Čejka J, Majzlan J, Sejkora J, Machovič V, Talla D (2011) Běhounekite, $U(SO_4)_2(H_2O)_4$, from Jáchymov (St Joachimsthal), Czech Republic: the first natural U^{4+} sulphate. Mineral Mag 75:2739–2753

Plášil J, Fejfarová K, Hloušek J, Škoda R, Novák M, Sejkora J, Čejka J, Dušek M, Veselovský F, Ondruš P, Majzlan J, Mrázek Z (2013a) Štěpite, $U(AsO_3OH)_2 \cdot 4H_2O$, from Jáchymov, Czech Republic: the first natural arsenate of tetravalent uranium. Mineral Mag 77:137–152

Plášil J, Hloušek J, Škoda R, Novák M, Sejkora J, Čejka J, Veselovský F, Majzlan J (2013b) Vysokyite, $U^{4+}[AsO_2(OH)_2]4 \cdot 4H_2O$, a new mineral from Jáchymov, Czech Republic. Mineral Mag 77:3055–3066

Scharmova M, Scharm B (1994) Rhabdophane group minerals in the uranium ore district of northern Bohemia (Czech Republic). J Czech Geol Soc 39(4):267–280
Semenov EI (1963) Mineralogy of rare earths. Publishing House AN USSR. 412 p. [in Russian]
Semenov EI (2000) Mineralization and ore-mineralization of rare earths, thorium and uranium. GEOS, 306 p. [in Russian]
Sidorenko GA (1978) Crystal chemistry of the uranium minerals. Atomizdat, Moscow. [in Russian]
Teterin YA, Kulakov VM, Baev AS, Nevzorov NB, Melnikov IV, Streltsov VA, Mashirov LG, Suglobov DN, Zelenkov AG (1981) A study of synthetic and natural uranium oxides by X-ray photoelectron spectroscopy. Phys Chem Miner 7(4):151–158
Urusov VS (2001) Genetic crystal chemistry: problems and solutions. Crystallogenesis and mineralogy. In: Inter. Conf. Proc., St.-Petersburg, p 405. [in Russian]
Vasilyev EK, Kashaeva GM (1974) Ushchapovskaya 3. F. Roentgen-metric determinant of minerals (phosphate class). Nauka, 206 p. [in Russian]
Velichkin VI, Solodov IN, Doynikova OA, et al (2009a) Study of geochemical and mineralogical features uranium ores of the Khokhlovskoye deposit. [in Russian]
Velichkin VI, Solodov IN, Doinikova OA, Nesterova MV, Chertok MB, Shugina GA, Shulik LS, Ganina NI, Rekun ML (2009b) Study of geochemical and mineralogical features of uranium ores of the Khokhlovskoye deposit. Materials on geology of deposits uranium, rare and rare-earth metals. Inform. Digest KNTS, vol 154. VIMS, Moscow, pp 69–93. (in Russian)
Vinichenko PV (2004) Theory of biogenic ore formation on example of uranium deposits. Irkutsk, Sosnovgeologiya, p 215. [in Russian]
Volkov NI, Dymkov YM, Doynikova OA, Dikov YP (2005) Mineralogical features of uranium ores of the Khokhlovskoye Deposit (Trans-Ural) prepared for ISL, vol 1. Uranium. Ore and Metals, Moscow, pp 60–70. [in Russian]

Chapter 5
Paleochannel Sandstone-Type Uranium Deposits of Vitim Ore Region

Further in this subsection present our results obtained from last-years studies uranium ores from seven paleochannel-type deposits of Khiagda Ore Field (KhOF) at Vitim uranium ore region in Buryatia, Russia. Here predominant ore component is ningyoite (U^{4+}-Ca-phosphate). These ores, little-known previously and unknown in Russia, represent a new type of commercial uranium ores—phosphatic black's one (Doynikova 2012). The high dispersity of the U-ore substance led to the choice of analytical electron microscopy (SEM, TEM) as the main research method.

Information Backstory Sandstone-type uranium deposits dominate world uranium mining. They are most cost-effective for in situ recovery. In 2015, more than half world uranium production (about 55%) was extracted from deposits associated with sandstones (Uranium 2016). These exogenous ores are formed as a result of sediments stratal−/ground—oxidation by infiltration of the oxygen-rich groundwater. In accordance with the IAEA classification (adopted in 2013), among the sandstone associated uranium deposits, three subtypes are most widely represented: *tabular*—Czech Republic (Strazh), Bulgaria (Marishsky ore field); *roll-front*—Kazakhstan (ChuSarysu type); *basal channel*—Japan (Ningyo-Togo, Tono), Canada (Blizzard, British Columbia), Argentina (Mendoza), etc.

In Russia, two of three currently operating uranium mines—Dalur (Dalmatovsky district, Trans-Ural region) and Khiagda (Buryatia, Trans-Baikal region) exploit the sandstone-type deposits (basal channel subtype). Industrial exploitation at these enterprises is carried out with use of the environmentally friendly method of in-situ leaching (ISL).

Joint stock company (JSC) "Dalur" conducts operation and development of uranium deposits in the Trans-Ural region with the average reserves: Dalmatovskoye (12 kt U), Khokhlovskoye (11 ktU), and Dobrovolnoye (8 ktU) (Uranium... 2010). These deposits are located in the marginal part of the West Siberian platform. In 2017, the uranium annual production was 592 tU. Currently, the bulk of uranium is mined at the Dalmatovskoye deposit. In 2017, Dalur mine has started preparations for the development of a new uranium deposit Dobrovolnoye. In addition, small

O. A. Doynikova, *Uranous Mineralogy of Hypergene Reduction Region*,
Springer Mineralogy, https://doi.org/10.1007/978-3-030-67183-9_5

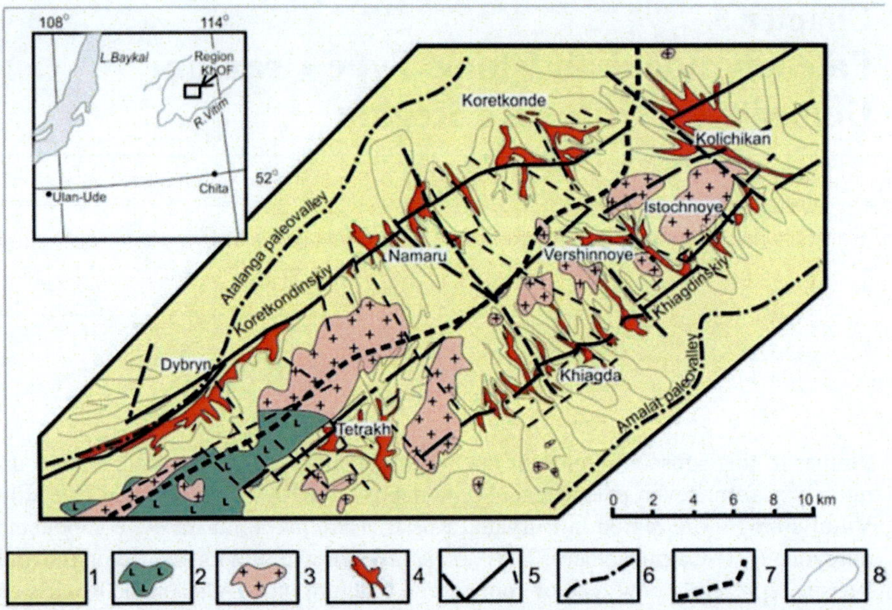

Fig. 5.1 Khiagda Ore Field (by [Uranium… 2010] with additions): 1–2—N-Q sediments: (1) terrigenous and volcanogenic-sedimentary, (2) flood-basalt; 3—basement rocks (Vitimkan granite complex); 4—projection of uranium deposits to the surface; 5—faults various in order; 6—paleovalleys axes (first order); 7—axis of the Baysykhan uplift; 8—paleo-ravines contours

deposits and ore occurrences of the same type are known in Mariinsky region of the Trans-Ural region: Malinovskoye, Novoye, Prigorodnoye, Ledyashevskoye and others (Tarasov et al. 2018).

JSC "Khiagda" conducts industrial operation and exploration of uranium deposits of the Vitim uranium-ore district. The main resources are hosted in the Khiagda Ore Filed (KhOF) which is located on the Vitim plateau (Vitim River, North-Eastern Trans-Baikal region). KhOF is a group of 8 paleovalley uranium deposits of Neogene age: Dybryn, Namaru, Koretkonde, Tetrakh, Khiagda, Vershinnoye, Istochnoye and Kolichikan. They were discovered in the 80-ies of XX century. Khiagda deposit was the first studied and it is why the entire Ore Field was named after Khiagda. All KhOF deposits are localized in paleovalleys embedded in crystalline basement (Fig. 5.1).

In addition, several spatially isolated small deposits and occurrences of the same type (Djilindinsky and others) are located in the northern and southern parts of the Vitim region, outside the boundary of the KhOF. Current exploration works take place in the northern part of the region. Another sandstone basal channel type deposit Balkovskoye (resources 15 ktU) is also known in the Miocene paleovalley sediments of the Don river ("Paleo Don") in Kalmykia, South Russia region (Uranium… 2010).

According to the IAEA classification (Uranium 2016), all the above-mentioned Russian deposits belong to the sandstone economic type (basal channel subtype). Russian geologists commonly call such deposits as paleovalley, paleo-riverbed or paleo-streambed. Dispersed friable black uranium ore mineralization of such deposits, hosted in the unlithified sediments, is composed of U^{4+}- minerals: uraninite, coffinite, ningyoite. As noting higher, uranous U^{4+} minerals here can occur in different proportions.

In Russian geological terminology such black U-mineralization is called uranium "cherny"; in foreign literature usually—«sooty uraninite». Our results of mineralogical AEM study of uranium ore composition in Vitim paleovalley-type deposits is a core of this subsection.

Detailed geological exploration of KhOF deposits has begun in 2009 by subsidiary of uranium holding company Atomredmetzoloto (ARMZ), which is a part of Rosatom, the Russian state corporation. Most of the exploration and drilling work was carried out by Rusburmash, exploration Service Company of ARMZ. In 2014, the main exploration and resources estimation for the Khiagda group of uranium deposits was completed (Uranium 2016). Published recently (Kochkin et al. 2014; 2017a; Tarasov et al. 2018) results of complex geological study of KhOF deposits have been told the total estimated KhOF resources are >45 Kt U. In 2017, annual U production was 693 t; production of 1kt U annually is expected by 2019 (Uranium 2016). Currently, Khiagda and Istochnoye deposits are under mining and Vershinnoye is preparing for development. During this period we have collected a unique U-ore material which study results are given below.

5.1 Paleochannel Deposits of Khiagda Ore Field, Vitim Plateau, Russia

Powdered fine dispersed black uranium ore mineralization of KhOF was considered as pitchblende-coffinite until starting detailed exploration from 2009 when AEM has been involved in the practice of mineralogical study. Our first stage evaluation of Khiagda black U-mineralization by ATEM method in early 1980s has already shown prevalence of U^{4+}-Ca-phosphate, ningyoite. However, this mineral was described as subordinate in the composition of predominantly nasturane-coffinite ores by other researches for this and other deposits of KhOF until recently. The first reports about phosphate composition of these U-ores were published in Tarkhanova et al. 2014 (Doynikova et al.; Tarkhanova et al). The results of our today study have confirmed the fine dispersity (micron dimensions) of black uranium mineralization and prevalence of phosphate in its composition which is typical for all deposits of KhOF. The first results of our detail study have discovered the new for Russia type of commercial U mineralization—phosphate blacks' ores (Doynikova et al. 2014). Like all uranium deposits with ningyoite mineralization (Doinikova 2007), Vitim deposits belong to the infiltration type deposits: of stratal—and ground/subsoil—filtration, according to classification adopted in (Kislyakov and Shchetochkin 2000).

Geological Background Khiagda ore field (KhOF) is located at the left bank of the Vitim River, within the Baysykhan uplift (Amalat basalt plateau) which can be traced by the granitoid basement outcrops in the field of the basalt cover (Fig. 5.1). The location of ore deposits is controlled by large paleochannels and their inflows, which are embedded into crystalline basement at the uplift slopes and are overlain by thick basalt cover.

The common geological features for all KhOF deposits are following: conditions of black U mineralization localization; lithologic and facial composition of U-bearing rocks, mineral composition of ores. The difference is only the size of paleochannels and of the ore bodies. Uranium black mineralization at all deposits is localized mainly in volcanic-sedimentary and terrigenous sediments, less often in overlying basalts and basement granitoids. Basement rocks are characterized by anomalous uranium contents (average 5.5–6.0 × 10^{-4}). Most researchers consider such rocks as source of uranium for these deposits (Uranium… 2010).

Uranium-bearing rocks are Neogene—Quaternary sediments of the Djilinda suite. It is deluvial-proluvial brownish gravel, gray-colored sand-clay sediments, fluvial sands, lacustrine-bog clayey siltstone, clay and peat. The rocks are enriched with carbonaceous organic matter in the form of plant remains and fine dispersed detritus. The average C_{org} content in the sediments is 0.1–0.3%, but sometimes locally is up to first percents. The admixture of volcanogenic material is in the upper part of sedimentary strata and contains ash, lapilli and volcanic debris.

Sedimentary rocks in the entire area are almost completely covered by thick (up to 200 m) layer of basalts and tuff-basalts (Amalatskoje plateau). There are some separate outcrops of basement granitoids at the western part of the KhOF.

Classical regularities of the roll-front ore bodies' formation at the redox boundary in permeable sedimentary strata have been observed on KhOF, as in the all infiltration sandstone-type uranium deposits. This is the boundary of unaltered gray-colored rocks with epigenetically modified, whitewashed, ones. (The whitish color of former gray-colored rocks is the result of epigenetic redox processes transit through them). It is assumed that ore mineralization was formed at the geochemical reduction barrier, which was created by bacterial decomposition of host rocks organic matter. The position of this boundary and its morphology in the section and in the plan is determined by the lithological composition of uranium-bearing rocks, their effective porosity and permeability.

Ore bodies in terrigenous and volcanogenic-sedimentary rocks are usually located in primary gray-colored rocks near to contact with the whitewashed ones, less often in the whitewashed rocks. The ore bodies of lenticular and more complex forms are localized in the axial parts of paleochannels. There is a prevalence of ore bodies of ribbon-shaped form in plan and of lenticular one in cross section (thickness n ÷ 20 m). They are jointing into ore accumulations (from n·100 m up to 8 km). More rarely ore bodies have irregular columnar shape and can be traced through the whole thickness of sedimentary strata partly extending into the basement rocks and overlapping basalts. It is assumed that localization of such bodies is caused by the carbon dioxide inflow from the basement into the ore-hosting layer along permeable zones of large faults (Kochkin et al. 2014; 2017b).

5.2 Composition of Uranium Ores on Khiagda Ore Field Deposits

Objects and Methods of Study Mineralogical data on the uranium mineralization of KhOF deposits—Dybryn, Namaru, Koretkonde, Khiagda, Vershinnoye, Istochnoye, and Kolichikan—have been received. The drill core samples with highest U grade from different lithological varieties were studied; the terrigenous sedimentary rocks are predominating among it. At the first six deposits have been studied samples of all varieties of ore-hosting rocks: light and dark sands, siltstones, granites, basalts. At the last one—only few samples from sandstone.

The studied ore samples from terrigenous sedimentary rocks are a loose material of weakly cemented sandstone or siltstone. Their gray and dark-gray color is due to carbonaceous organic. Effusive rocks with uranium mineralization are represented by altered porous basalts and their tuffs. U-mineralized samples from the basement are weathered granites. In basalts and granites uranium mineralization fills cracks and voids and is also hosted in veins with chlorite and clay minerals.

Composition of uranium ore substance was studied predominantly by AEM, and radiography has been used as preview method; then by AEM equipments (JSM-5610LV + INCA-450, 25 kV; JEM-2100 + IETEM INCA-250, 100 kV). Diagnostics of U-ore minerals was carried out by complex methods of TEM and SEM (SAED, EDS, BSE-imaging). Samples were studied in different EM preparations: handpicked individual grains, transparent polished sections, and polished sections.

Ores' Mineralization of KhOF Deposits The overall balance of uranium in the sandstone uranium ores is composed of uranium minerals, accessory and sorbent minerals. In all the deposits studied on KhOF, uranium mineralization is of the same type. Visually, by morphology, U-mineralization is characterized as uranium blacks (*cherny*). The uranium-bearing rocks are various but the characteristics of ore substance from KhOF deposits are general: high dispersity and cementing nature of ore deposition, their mineral composition which is manifested in all kinds of host rocks. More than 50% of uranium mineralization in the sedimentary rocks is confined to clay cement of sandstones, and less forms coatings on clastic rock-forming minerals, according to the radiography data. In sands uranium minerals form rims and coatings on grains (Figs. 5.2 and 5.3), and develop along the cracks of altered feldspars. Mineralization of higher grade is related to carbonaceous plants remains.

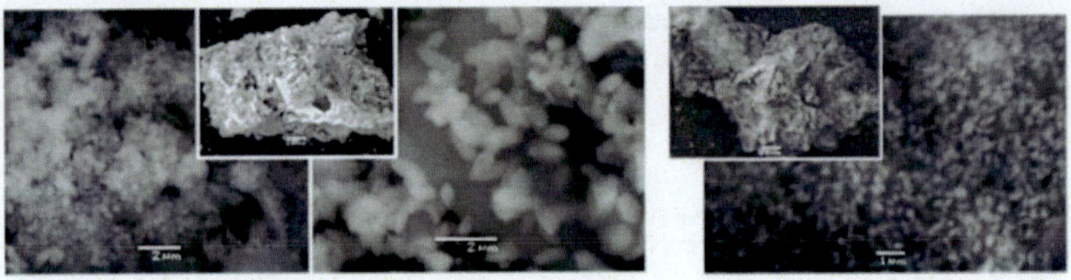

Fig. 5.2 Ore sand grains (insets) and their enlarged surfaces: (**a**) crusts and aggregates of spindle-shaped ningyoite crystals (light, white), (**b**) needle-shaped submicron ningyoite in Al-Si mass. Here and below—BSE images

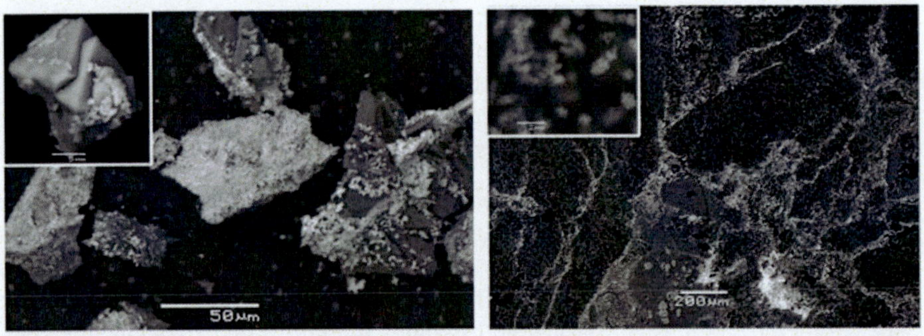

Fig. 5.3 The forms of uranium mineralization precipitation (bright, light): (**a**) the rich ore, granites,—ningyoite crystals on the surface of mineral aggregates and grains, on pyrite surface (inset); (**b**) basalt—ningyoite-sulphide formations (light) show the solutions paths; inset—submicron ningyoite crystals in the vitreous material, from a wide "flow" in the center (thin section)

Uranium mineralization is represented by the crystalline and mineraloid form of ningyoite. The phosphate form of tetravalent uranium significantly dominates in the composition of the ores. The main ore mineral was diagnosed as ningyoite CaU $(PO_4)_2 \cdot 1\text{-}2H_2O$ (rhabdophane group) and its isomorphic varieties containing Fe and S in all types of host rocks. Minerals with idealized formula $(Ca,U,Fe) \cdot (P,S)O_4]$ $\cdot nH_2O$ are predominant.

For the *crystalline* form of this phosphate main characteristic is the micron size of crystals (n·μm) (Fig. 5.4). The size of massive crystalline ningyoite accumulations rarely reaches n·100 μm. The largest crystals of ningyoite (7–8 microns) with a pyrite rim on the faces were found in the micro cavities of large carbonaceous fragment from clay interlayer (Fig. 5.5). Previously unknown reflectivity of ningyoite was detected here and was defined as being similar to quartz values. Submicron ningyoite crystals (elongated shape) were determined in the cementing Al-Si-mass of uranium-bearing sands (kaolinite—hydromica material) and in carbonaceous debris (Fig. 5.2b). Ningyoite often contains Ce, La, and Nd, less often the impurities of Sr, Y, Zr and Ti. Large crystals contain Ce 2%, La 1%, and Sr 1%.

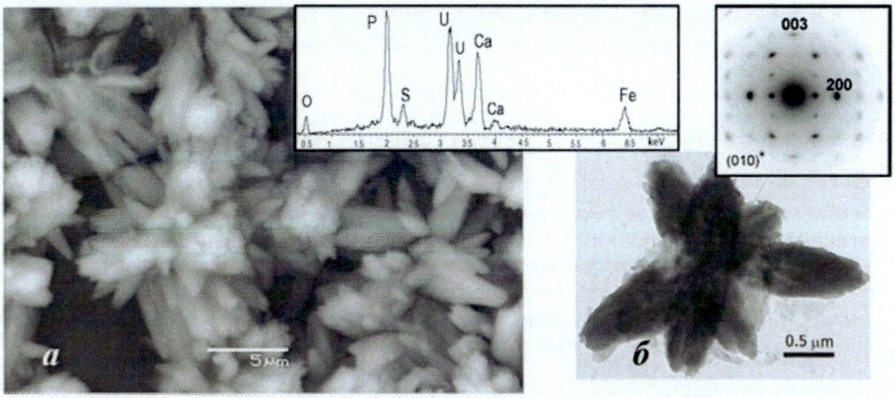

Fig. 5.4 Typical diagnostic characteristic of ningyoite under EM: (**a**) crystals intergrowths, SEM image, and composition spectrum, (**b**) TEM image and characteristic SAED pattern, plane (010)*

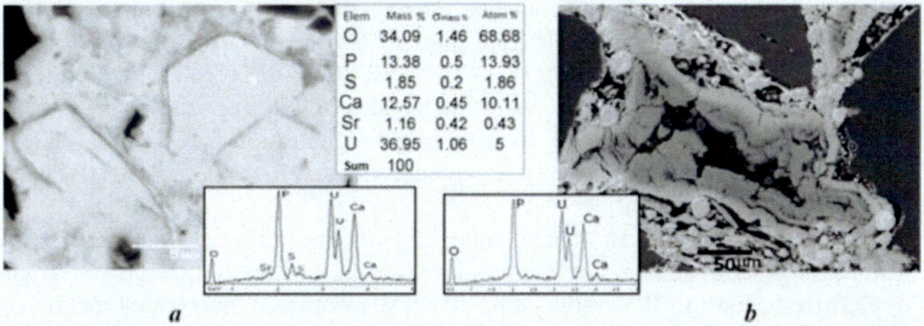

Fig. 5.5 Mineraloid form of ningyoite: (**a**) U-ore mass with a sulfide component (light-gray) with large ningyoite crystals (bright) within; the composition spectrum and analysis result of the crystals; (**b**) cement in the U-rich sands—mineraloid ningyoite and its spectral composition (the bright outer layer); association with pyrite framboids (rounded)

Mineraloid form of ningyoite as solid uranium mineral gel was often observed in the studied samples. The morphology of such secretions has curved character that indicates the primary plasticity of mineral material (Fig. 5.5b). This type generates independent forms or phyto—and pseudomorphoses. But more often the mineralization is represented as formless mineraloid aggregates of slightly differentiated mineral mass. The size of these segregations rarely reaches n·10 μm. Their composition U-P-Ca-Fe-S (often with Al and Si) is usually inhomogeneous. Spots with different ratios of ningyoite and pyrite components are represented in the colloform mineral mass without visible phase boundaries. The mineraloid form of the "pure" (without impurities) ningyoite is rare exception. Analysis of relatively large mineraloid forms (without contribution of associated sulfide minerals) shows the relative constancy of the "sulphide component" in their composition. This confirms

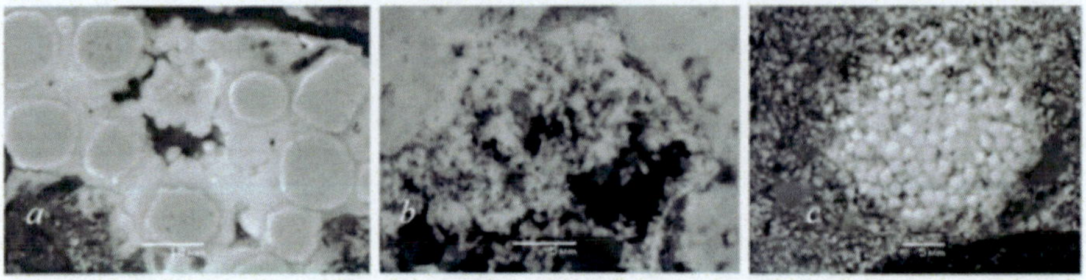

Fig. 5.6 Ningyoite-pyrite associations: (**a**) crystalline ningyoite mass (light) around the rounded pyrite framboids (dark-gray); (**b**) accumulations and crusts of ningyoite microcrystals in the space among pyrite; (**c**) framboid from cube pyrite crystals (bright) with ningyoite cement (light gray); the impregnation of submicron ningyoite around

the predominance of the Fe-S-containing varieties of ningyoite. The contribution of Al-Si component to the mineraloid composition is always different.

REE-ningyoite occurs in all types of ore-bearing rocks (sands, granites, basalts). It appears as aggregates of needle-shaped microcrystals (0.1n ÷ n μm) or sulphide-ningyoite mass. The amount of REE reaches 10%; sometimes up to 1% Sr is noted. In ningyoite crystals Σ TR ≈ 2 ÷ 13 wt% (Ce—max 6%); the TR ratio: Ce > Y > La > Nd > Pr. The largest aggregates of TR-ningyoite crystals are n • 10 μm in size (Dybryn deposit). Its determined composition is $(Ca_{0.67}U_{0.25}TR_{0.08})PO_4$. The idealized formula of TR-ningyoite is (Ca, TR, U) $[PO_4] \cdot nH_2O$, where TR = Ce, La, Nd.

U- metacolloid with Th content (ningyoite-brockite material) was identified in ore-bearing sand among altered products of dark-colored minerals of granite (chloritized biotite). The composition of such material is ningyoite-brockite-sulphide. Similar phases of more complex composition are also present in weathered granite (pebbles in sandstone) in the form of inclusive nodular accumulations (n·10 μm).

All uranium minerals closely associate with sulfides framboides (Fig. 5.6).

Besides U-mineraloid and U-crystal forms, *sorbed form* of uranium was identified in U-bearing samples with abundant goethite cement. Thorough search of U-phases in such ore samples by AEM-methods was unsuccessful, despite of their high U grade. The presence of sorbed uranium in goethite has been revealed only from the data of digital autoradiography (alpha detection) (Fig. 5.7). Besides of iron hydroxides, U is probably present in clay and carbonaceous materials in the same form. The presence of this sorbed U form in the host rocks is manifested as a formation of yellow incrustations (uranyl phases) on the surface of the drill core samples after drying.

Uranyl mineralization occurs in sharply subordinate amounts. Mineral particles have distinct laminated or fibrous morphology (Fig. 5.8). In the composition spectra of such particles, the ratio of peaks is U > P > > Ca. The mineral was preliminary identified as skupite by the fibrous morphology. Possible presence of autunite admixture in samples is allowed.

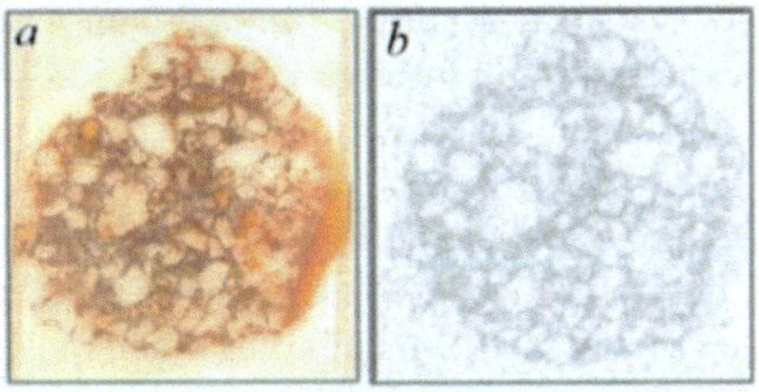

Fig. 5.7 (**a**) The thin section of U-ore sample with goethite cement, (**b**) its digital autoradiography (α-detection), showing the presence of sorbed uranium

Elem	Mass %	σ mass %	Atom %
Al	0.78	0.19	1.32
Si	1.87	0.17	3.03
P	2.6	0.26	3.83
S	3.22	0.19	4.58
Ca	2.41	0.18	2.74
Ti	1.92	0.2	1.02
Fe	3.75	0.24	3.06
Zr	3.03	0.61	1.51
U	56.86	0.77	10.9
O	23.57	0.65	67.19
Sum	100		

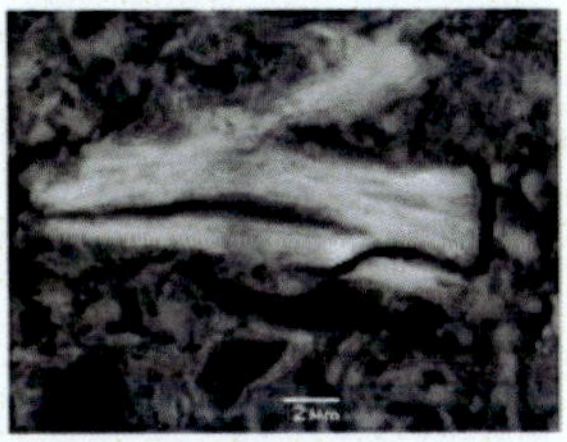

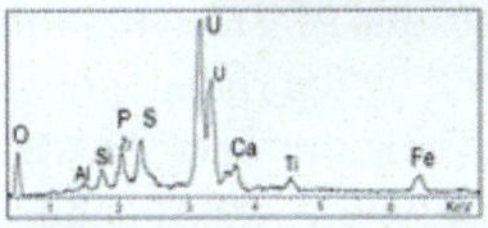

Fig. 5.8 The uranyl-mineral crystal (light) with pronounced layered-fibrous morphology; result of analysis and composition spectrum

Fig. 5.9 Ningyoite microcrystals (bright points) in sagenitic lattice of the altered titanate grain, large lamellas—titanomagnetite (?)

U-containing mineralization is often found on leucoxenized titanate grains in the form of colloform aggregates or crusts (Chap.4, Fig. 4.11). It is U-containing Ti-oxide phase (hydro-anatase?) with the variable amounts of elements (wt %): up to 2% U, up to 4% Fe, 1–2% P, 0.1–0.3% Ca, and less often 1.5–2 wt % Zr (in the case of presence of clastic zircon in sands). Such fixation of U, Ca and P in amorphous phase (Fig. 5.8a) is considered as initial stage of uranium-mineral formation (phosphate?). The same content of the admixtures of "ore" components U, Ca, P (≈ 1% wt) is determined quite constantly in composition of altered zircon grains. In the altered titanate grains rarely fixed tiny uranous mineralization microcrystals (Fig. 5.9).

Associated mineralization. Our research has identified the presence of *REE mineralization* in ores of seven KhOF deposits. By-product extraction of REE from industrial solutions reduces the production cost of uranium mining. Such mineralization is clearly manifested in all types of ore-bearing rocks. There are three types of REE concentrations: (i) REE minerals—monazite $CePO_4$ and xenotime YPO_4 (clastic grains of accessory minerals of granite), (ii) new formations of supergene TR-minerals as rhabdophane, (monazite?), (iii) notable REE admixture in the U-mineral phases.

The clastic monazite grains (up to n·10 μm) in uranium enriched samples are often cracked with signs of dissolution; they are altered to varying degrees. The grains of the primary accessory monazite have signs of later changes with gain of Ca, Th and U (up to 10%). The less common smaller grains of xenotime (n μm), have higher degree of their preservation. According to Perelman (1968, Perelman and Kasimov 1999, Hydrogenic… 1980), such different sustainability of Ce- and Y-phosphates indicates a high acid environment of alteration processes in which LREE +Y are mobile and HREE do not migrate.

Rare inclusions of *TR-containing* secondary minerals (later generation) in sands are represented by the rhabdophane group minerals in the form of fine impregnation (Fig. 5.10). Their presence in the KhOF ores indicates the loss of REE from the basement granitoid rocks (denudated in headstreams of paleovalleys). REEs were transported by uranium-bearing solutions, same as U, and deposited under conditions close to ningyoite formation.

Uranium minerals closely associate with sulfides of Fe, Zn, (Cu). It should be noted that ningyoite is developed preferably in spots with abundant pyritization (Fig. 5.6). Framboids of pyrite, pyrite-greigite, marcasite, and pyrite-marcasite are common. It is a peculiar prospecting indicator of U-phase under EM. Newly formed idiomorphic microcrystals of pyrite, marcasite, greigite, sphalerite, galena, and pyrrhotite are found. In addition, yordizit (colloform molybdenite variety), ilmenite, rutile, galena, fluorine-apatite, zircon, titanite, native copper and Cu-Sn intermetallides are detected in the composition of uranium ores.

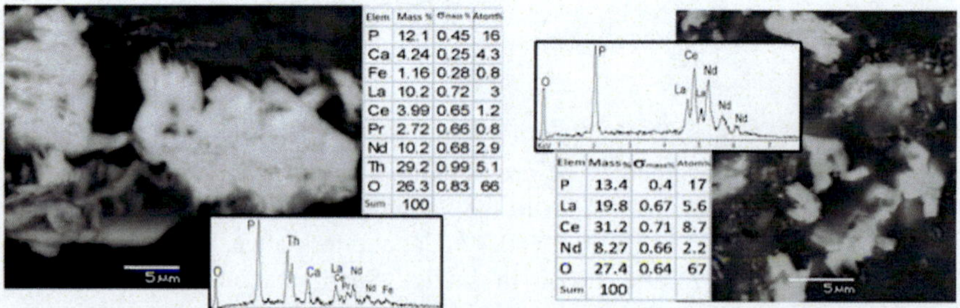

Elem	Mass %	σmass %	Atom%
P	12.1	0.45	16
Ca	4.24	0.25	4.3
Fe	1.16	0.28	0.8
La	10.2	0.72	3
Ce	3.99	0.65	1.2
Pr	2.72	0.66	0.8
Nd	10.2	0.68	2.9
Th	29.2	0.99	5.1
O	26.3	0.83	66
Sum	100		

Elem	Mass %	σmass %	Atom%
P	13.4	0.4	17
La	19.8	0.67	5.6
Ce	31.2	0.71	8.7
Nd	8.27	0.66	2.2
O	27.4	0.64	67
Sum	100		

a *b*

Fig. 5.10 New formations of TR-minerals: (**a**) REE-rhabdophane, crystals (bright) in the inter-granular cavity of granite pebble from sand, analysis result and composition spectrum; (**b**) rhabdophane (white) in a quartz-carbonate mass, analysis result and composition spectrum/sample of near-ore siderite

Summarizing results on mineralogical study of U-ores in seven deposits from KhOF Our studies of U-ores in paleochannel deposits from Vitim district have confirmed discovering here the new for Russia type of commercial uranium ores—phosphate blacks ones. It was shown next mineralogical features of ningyoite: its isomorphic Fe-S-containing phase; micron size (in crystalline and mineraloid forms); it reveals as fine dispersed cement in sandstone and closely associated with pyrite (often as framboids). We do not exclude the principle possibility of presence of U^{+4} oxide and silicate forms, which we could not find in KhOF ores (perhaps because the richest specimens were studied).

The mineralogical study of the KhOF uranium-phosphate ores has confirmed their deposition same as in all known sandstone-hosted U deposits (Hydrogenic ... 1980; Perelman and Kasimov 1999); namely, their deposition from oxygen-containing U^{6+}-bearing water due to U reduction and precipitation on reduction barrier. Oxidation of organic matter creates favorable environment for the reproduction of anaerobic microorganisms that continue to oxidize organics (after aerobic ones) and provide a reductive medium like as in other infiltration sandstone U deposits. The roll forms of ore bodies at KhOF emphasize the general features of ore deposition in the sandstone-type U-deposits.

When clarifying the genesis of uranous phosphate mineralization, the question of the phosphorus source is primarily considered. As noted above and more detail discussing further, it can't be clastic apatite: there are no traces of dissolution of its grains, which associate with ningyoite. Results of modern research in the fields of ecology and environmental mineralogy have helped the author to solve this issue. The analysis of deferent publications was revealed that phosphorus source in natural solutions is the decomposition products of organic matter (plant residues) (Doynikova 2016, 2017). The next Chap. 6 is devoted to the biogenic aspect of uranous minerals formation in sandstone-type deposits and discusses these questions more detail.

5.3 Conclusions

- In the basal-channel (paleovalley) uranium deposits of the KhOF (Vitim U-ore region), the ores are composed of phosphate-black uranous (U^{4+}) mineralization that manifested in all varieties of ore-hosting rocks (sand, overlying basalts, underlying granites).
- In all KhOF U-deposits, the ores are dominated by fine dispersed phosphate ningyoite (in crystalline and mineraloid forms) and its isomorphic Fe-S-containing phase. Uranium mineralization is represented as cement and closely associated with pyrite (often in the form of framboids).
- The phosphate compositions of black uranium ores, as well as the presence of biologically active lacustrine-bog deposits in the ore strata, indicate the

predominant control of biogenic factor during formation of uranium deposits of the basal-channel (paleovalley) type.

References

Doinikova OA (2007) Uranium deposits with a new phosphate type of blacks. Geol Ore Deposit 49 (1):80–86. https://doi.org/10.1134/S1075701507010047

Doynikova OA (2012) Uranium mineralogy in Hypergenesis reduction zone (according to electron microscopy data). Fizmatlit, Moscow, p 216. [in Russian]

Doynikova OA (2016) Phosphate composition of uranium sooty (blacks) as indicator of their biogenic nature. Proc higher educational establishments. Geol & Exploration 5:17–29. (in Russian with English abstract). https://doi.org/10.32454/0016-7762-2016-5-17-29

Doynikova OA (2017) The biogenic aspect of the uranium blacks formation (manifestation of the law named after V.I. Vernadsky). In: Kasimov N, Gennadiev A (eds) Landscape geochemistry / 100th anniversary of AI Perelman. Publ H APR, Moscow, pp 524–543

Doynikova OA, Tarasov NN, Mokhov AV (2014) A new phosphatic type of uranium deposits in Russia. Dokl Earth Sci 457(2):910–914. https://doi.org/10.1134/S1028334X14080030

Hydrogenic uranium deposits (1980) Ed. Perelman AI. Atomizdat, Moscow: 270 (in Russian)

Kislyakov YM, Shchetochkin VN (2000) Hydrogenic ore formation. Geoinformark, Moscow 608. (in Russian)

Kochkin BT, Novgorodtsev AA, Tarasov NN, Martynenko VG (2014) Morphology of ore bodies and genesis of uranium deposits in the Khiagda ore field. Geol Ore Deposit 56(6):479–492

Kochkin BT, Solodov IN, Ganina NI, Rekun ML, Tarasov NN, Shugina GA, Shulik LS (2017a) Geochemical features of the ore-bearing medium in uranium deposits in the Khiagda ore field. Geol Ore Deposit 59(5):341–353

Kochkin BT, Tarasov NN, Andreeva OV, Nesterova MV, Golubev VN (2017b) Polygenetic and polychronic uranium mineralization at deposits of the Khiagda ore field, Buryatia. Geol Ore Deposit 59(2):141–155

Perelman AI (1968) Geochemistry of epigenetic processes (supergene zone). Publ H Nedra, Moscow 331. (in Russian)

Perelman A, Kasimov N (1999) Landscape geochemistry. Publ H Astrea-2000, Moscow, p 768. (in Russian)

Tarasov NN, Kochkin BT, Velichkin VI, Doynikova OA (2018) Deposits of the Khiagda uranium ore field, Buryatia: formation conditions and ore control factors. Geol Ore Deposit 60 (4):347–354

Tarkhanova GA, Dubinchuk VT, Chistyakova NI, Nikitina ES, Prokhorov DA, Nechelyustov GN, Ruzhitskii VV (2014) Features of mineral composition and conditions of formation of ores of the Vershinnoye deposit. Razved Okhr Nedr Exploration and protection of earth interior 6:7–13. (in Russian with English abstract)

Uranium (2016) Resources, production and demand. OECD (2016) NEA/IAEA/ 7301

Uranium of the Russian's Interiors (2010) Ed Mashkovtsev GA VIMS Moscow: 850 (in Russian)

Chapter 6
Nature of Uranous Ore Formation in Hypergenesis Region

The most important contribution of sandstone-type U-deposits with black ores to the world's uranium mining explains the increased interest in finding out the genesis of such ores. The biogenic aspect of uranous ore formation is considered further in light of doctrine of geochemical barriers and modern data of geomicrobiology and environmental mineralogy (of their' results on study products of bacterial activity).

Some Fundamental Background Information
As mentioned above, uranium blacks (high-dispersed uranous mineralization) are formed under reducing conditions during activity of redox weathering processes. Uranium black ores are characteristic for infiltration deposits in sedimentary cover; they are manifested in almost all uranium deposits in hypergenesis region. Localization of this region in the Earth biosphere is key point in discussion of such U-ore genesis.

The biosphere includes soils, weather crusts, groundwaters and the deeper horizons of the Earth. To date, microorganisms have been found at depths of hundreds and several thousand meters from the Earth's surface. Their active participation in almost all geological processes forming sedimentary cover of the planet has been established over last decades. The bacterial nature of a number of iron, manganese, gold and copper deposits formation has been established (Zammit et al. 2015; Markov 2015).

Consideration of uranous minerals formation in sedimentary strata is impossible without attraction of the basics of landscapes geochemistry and teaching about geochemical barriers. The teaching on landscapes geochemistry, based on the views of famous Scientifics V.I. Vernadsky, A.E. Fersman and V.M. Goldschmidt, was developed in the 40s by B.B. Polynov and was expanded in detail by his pupils A.I. Perelman and M.A. Glazovskaya in independent scientific direction, which "allows analyzing many processes occurring in weathering crusts, artesian basins, and other biocos systems" (Perelman and Kasimov 1999). Landscape geochemistry was rapidly developing in the former Soviet Union (USSR) largely due to A.I. Perelman works on study a number of sandstone-type uranium deposits in

O. A. Doynikova, *Uranous Mineralogy of Hypergene Reduction Region*, Springer Mineralogy, https://doi.org/10.1007/978-3-030-67183-9_6

Central Asia. The foundation for the geochemistry of hypergenesis zone was laid by him (Hydrogenic uranium deposits 1980; Perelman 1968; Perelman and Kasimov 1999). It was shown that in tundra, taiga, humid tropics and deserts one should look for ore deposits in different ways, taking into account geochemical features of these landscapes.

In A.I. Perelman's works soil formation caused by bacteria activity is considered as model of weathering and crust formation processes. He repeatedly emphasized that *redox* zoning in soils is a model of more grandiose geological processes taking place in biocos systems (Perelman and Kasimov 1999). The concept of biocos systems was formulated by the famous Russian scientist V.I. Vernadsky, developing the ideas of his teacher V.V. Dokuchayev.

Biocos systems are natural formations that appeared with the participation of living matter in process of evolution of Earth's envelope. These natural systems (e.g. soil) consist of inert and living bodies and are natural components of the biosphere. Term is often used as synonym for *ecosystems.* V.I.Vernadsky in his works for the first time showed the total effect of living matter activity in geological history. He noted that all *physical and chemical properties of* biocos systems require—sometimes extremely large—corrections, if their study does not take into account the manifestation of living matter into them (results of its activities). The followers of Vernadsky noted the outstanding role of organic (bacterial) world during diagenesis of sediments, for example it was famous Russian lithologist N.M. Strakhov (1957). However, according to A.I. Perelman, Vernadsky's biogeochemical ideas have gained general recognition only in recent decades due to the acute, emerging environmental problem. Landscape geochemistry has become one of the theoretical foundations for solving various environmental problems.

A.I. Perelman showed that the main factor of element migration in the Earth's crust is *biogenic migration* and formulated next law on geochemical activity of organisms. "Migration of chemical elements in biosphere is carried out either with direct participation of living matter (biogenic migration), or it occurs in environment whose geochemical features (O_2, CO_2, H_2S, etc.) are due to living matter, both that which currently inhabits the biosphere and that which has been active on Earth throughout geological history". A.I. Perelman proposed to call this provision Vernadsky's law, considering it to be one of the basic laws of geochemistry.

For further consideration of processes hypergenic uranium ore formation, taking into account role of living organisms and their participation in geological processes, another fundamental notion of geochemistry "geochemical barriers", formulated by A.I. Perelman, is necessary. *Geochemical barriers* are those parts of Earth's crust, where at a short distance there is sharp decrease in intensity migration of chemical elements and, consequently, their concentration take place. This concept is based on similarity processes of elements concentration in the Earth's crust: accumulation in landscapes and deposition from aqueous solutions.

When considering *biogenic migration*, A. I. Perelman chose as a basis processes common to all parts of biosphere—decomposition of organic substances. By his' works, in landscape geography knows that arid and humid conditions of organic matter accumulation differ in the results of mortmass accumulation (mass of dead organic matter in ecosystem). In *humid climate*, the accumulation of organic material and humus in soils is faster than its decomposition by microorganisms. Humid climate provides for the accumulation of phytomass in sediments. Its considerable reserves, which were destroyed by microorganisms but not up to whole decay (mortmass), have been accumulated here too. In comparison with sediments of *arid climate*, here is abundance of mortmass. Probably the above mentioned, organic substance previously processed, oxidized by aerobes (which serves as nutrient for anaerobe) should be correlated (identified?) with the mortmass in the ore-bearing strata.

If there is not enough heat, the decomposition of plant remains slows down, they do not have time to completely decompose, and the landscape accumulates excess mortmass. Mortmass reserves in arid areas are always smaller than the accumulated phytomass reserves. The rate of biodegradation in arid and humid conditions varies, depending on the species composition and the intensity of life of microorganisms. In addition in soil-science textbooks, the speed of humus formation is considered depending on the water-air regime and granulometric composition of soils. "In sandy and sandy loamy soils with good heating and aerated soils decomposition of organic residues is fast, a significant part of them is mineralized, humus substances not many, they are poorly fixed on the surface of sandy particles. Under equal conditions in clayey and loamy soils the process of decomposition organic remainders is slower (due to lack O_2). Humus substances are fixed on the surface of mineral particles and accumulate in the soil" (Marchik and Efremov 2006). It is such processes' results we observe in black U-ores.

Background History

Briefly formulating the theory of formation essence of so-called exogenous and epigenetic uranium deposits, Russian academician N.P. Laverov said: "The *biogenic factor manifests itself as a generator of reduction settings*". Russian uranium geologist's interest on biogenic features of uranium-bearing sedimentary strata was clearly manifested by to 70s of last century. Important role of microorganisms in formation of geochemical barriers that localize uranium accumulation was first declared by A.K. Lisitsin (1975). His conclusion about bacterial nature of high reduction potential, leading to uranium ores formation in oxygen-free environment, is the basis of classical views on creation of infiltration (exogenous-epigenetic) uranium deposits (Hydrogenic uranium deposits 1980; Rare metal uranium ore formation in sedimentary rocks 1995; Kislyakov and Shchetochkin 2000). Biogenic nature of pyrite, associated with U blacks' ores, was proved using isotope analysis of pyrite sulfur from U-ore body; from which assumption about biogenic origin of accompanying uranium mineralization was made (Lisitsin 1975). Subsequent publications on geochemistry of sandstone uranium deposits, was supporting these conclusions, only added anaerobic bacteria forms to the list, specify and detail them. However, until recently, Russian geologists considering biogenic processes (the activity of microorganisms) as introduced (attendant) ones, which only accompanying the formation of uranium deposits in sandstones. And until recently, it is *anaerobic* bacteria that were considered the main agents that creates sharply-reducing environment, and *aerobes*' role remained practically unnoticeable. As a rule, many authors limit role of bacteria to catalytic function, treating it as an "auxiliary" one, giving priority in uranium deposition to geochemical processes.

During study sandstone type uranium deposits Russian geologists was found that the most intensive biochemical processes occurring on borders of heterogeneous media, where maximums microbiota development is manifested (Hydrogenic uranium deposits 1980). It is in such places, on the *redox* boundary, black uranous mineralization is formed; i.e. uranium deposition on reducing geochemical barrier takes place. It is in such places uranous deposition in the hypergenesis region is emerging.

All uranium black ore minerals (oxide, silicate and phosphate forms) were formed in similar environment that allows us to generalize their formation conditions in all sandstone hosted uranium deposits.

Earlier, when studying conditions of uranous sandstone ores formation, domestic geologists established connection of significant part uranium with water-soluble organic matter (Lisitsin 1975). In the newest educational literature U^{+4}-mobility is marked mainly in form of uranium-organic complexes of fulvic acids (Perelman and Kasimov 1999). Connection of U^{4+} with organic matter of colloids under reducing conditions of soils and mountain workings is known (Wang et al. 2013). Colloidal transport of U^{4+} is established in pore waters of deposits (Malkovsky 2011; Malkovskii et al. 2014; Malkovsky et al. 2015; Wang et al. 2014); its colloidal polymers for very acidic environments are known in inorganic chemistry (Priyadarshini et al. 2014).

Since 1990s, in connection with the decision of ecological problems there are many foreign and domestic publications on bioremediation—removal of radionuclide contamination (including U) in soils, sediments and ground waters through using microbial activity (Lovley et al. 1991; Khizhnyak 2013; etc); the number of such publications has been increasing significantly in recent years. Using bacteria for U fixation in contaminated near surface environment is considered as an alternative to complex and costly recultivation technologies. The biogenic aspect of uranous minerals formation in sandstone-type deposits was revealed in the analysis of modern publications on geomicrobiology and environmental mineralogy (Doynikova 2016, 2017). Review a number of publications related to problems of radionuclides immobilization in near-surface environment (Lovley et al. 1991; Suzuki et al. 2005a, b; Sivaswamy et al. 2011; etc) has emphasized the main role of bacteria in radionuclides accumulation, when reduction (gley) environment is creation (i.e. reduction barrier).

Uranium immobilization products in natural environments are studied by ecological (environmental) mineralogy. Its' results are also important for ores development using microorganisms (Calas et al. 2015). Diagnostication of dispersed mineral forms of microbial-reduced uranium is performed using modern methods (EXAFS, HRTEM, XRD, EDS, etc.). Until early 2000s, in experiments on uranium bacterial deposition in microbial-saturated water environment, sediment composition (without diagnostics) was considered alternatively as tetravalent uranium oxide. Review of uranium microbial immobilization (Sivaswamy et al. 2011) shows that uranium fixation often takes phosphates form, less often uraninite. Along with uraninite and ningyoite, the non-crystalline ("non-mineral") form of uranium U^{+4} in the form of "biogenic nanoparticles" of UO_2 was detected (Singer et al. 2009). Biogenic uraninite already is studying in many investigates both microbiological and ecologic-biochemical orientation (Cerrato et al. 2013; Lezama-Pacheco et al. 2015). In recent decades, and mineralogists studying uranium ores in hypergenesis region are increasingly paying attention to important role of biogenic processes in U-ore formation. Results of modern research confirm leading role of biogenic processes that create reducing environment and providing ore formation.

Microorganisms are found at the depths n·100 m ÷ n·1000 m from the land surface. Today it is known that they are active in all geological processes forming a sedimentary Earth crust. Wide range metals release in bioleaching processes has been established. It was found that microbial catalyzed redox processes of metal redistribution in the Earth's crust can often produce economically significant metal enrichment. The bacterial nature of some Fe, Mn, Au, Cu deposits is already known (Zammit et al. 2015; Southam and Sanders 2005; Erlich and Newman 2008; Calas et al. 2015; etc). The sandstone-type uranium deposits are one of specific example of this kind ore formation, as a result of bacterial working. Such bacterial activity is considered by us as a factor of U-ore accumulation, which indicates the biogenic nature of uranium deposition in sandstones. Sandstone roll-front U-deposits in supergene systems, where weathering is stimulated by microbial activity, are considered as lateral development of hypergene processes from aerobic weathering to anaerobic enrichment (Zammit et al. 2015). Such bacterial activity is considered by us as a factor of U-ore accumulation, which indicates the biogenic nature of uranium deposition in sandstones. Black uranium ores are formed under reducing conditions at the front of redox weathering processes.

6.1 Biogenic Aspect of Uranous Ore Formation

Present-day microbiology considers the microbial community as a system of heterogeneous, functionally diverse organisms, interacting with each other (Zavarzin and Kolotilova 2001); anaerobic environment is considered as trophic structure of microbial community (when energy and substance of any organisms are consumed by other organisms).

Creation of anaerobic reducing medium is caused by the *summary* action of microbiological processes (by microorganisms' community) in the *integrated* redox barrier space, where aerobic and anaerobic microorganisms coexist. Anaerobes in sedimentary layers continue oxidation of organic matter, completing processes of its aerobic decomposition. Anaerobic oxidation is inextricably connected with the previous aerobic process; these aerobic-anaerobic oxidation processes are closely connected among themselves. Accordingly, geomicrobiology is based on the unity of microbial community in ecosystem. As a result, the oxidation zone is inherently integral part of the ore-forming barrier.

Such holistic view of aerobic-anaerobic bacteria community in processes of biogenic ore formation at uranium deposits is contained in the publication (Vinichenko 2004). Here noted present-day accumulation of uranium in peatlands and coal shale is accompanied by a local increase in uranium concentration, which is explained by uranium sorption/fixation by bacterial cells on the redox barrier. Marked in this publication local formation of rich ore-bearing solutions, in which "uranium content reaches hundreds of mg/liters", does not contradict known data on local accumulation of active reducing agents in solution, that causing a sharp decrease in Eh (Lisitsin 1975). V.P. Vinichenko considers the mechanism of such enrichment with uranium in the full range of redox barrier: he believing that it is *aerobic* bacteria, oxidizing organic matter that produce an abundance of organic acids and thus ensure emergence of a reductive environment, creating (preparing) the conditions for life of *anaerobes*.

Considering in general the process of biogenic black uranium ore formation, it is necessary to evaluate the activity of both anaerobic so previous (and coexisting) aerobic bacteria, understanding the unity of the microbial community in the ecosystem. Taking into account the activity of only anaerobic microbiota in the formation of uranium mineralization is not objective enough today.

The uranous-mineral environment occurs locally at the boundary of the redox conditions and moves with the rolling front. The process of formation black mineralization can be presented as follows. Decomposition of detritus leads to a local increase in acidity (organic acids) and, probably, a concomitant increase in uranium concentration in solution due to bacterial sorption i.e. fixation in this volume. Further activity of the bacterial environment (anaerobes begin to dominate aerobes) leads to mineral "sorption" (deposition) of reduced forms of uranium U^{+4} in the form of nano-sized "biogenic" particles on the surface of sand grains (Singer et al. 2009; Lezama-Pacheco et al. 2015). The activity of microbial community in the ore formation medium (it is reduction zone of the redox boundary) leads to further accumulation of U^{+4} with the formation of its mineral forms, which create a cementation mass in the intergranular space of sandstones. In this reducing environment, colloidal particles are the most important form of uranium transport (Kalmykov 2008; Malkovsky 2011; Malkovskii et al. 2014; Malkovsky et al. 2015; Wang et al. 2014; Priyadarshini et al. 2014).

Microbiologically caused, mechanism for moving reducing environment (jointly with redox barrier) can be appears as follows. Growth of aerobic bacteria (due to oxidation of detritus) creates a nutrient medium for anaerobes, thus ensuring movement of reduction zone (enriched with anaerobes) along sedimentary strata in direction of water filtration. Fading of anaerobic activity beyond ore deposition border (that reliably established by geologists) occurs probably as a result of deficiency nutrients that was prepared by aerobes. It is assumed that the movement of such joint biogeochemical redox barrier occurs as a result of expenditure of nutritious base for bacterial community (Doynikova 2016). This is manifested in the absence of organic matter (detritus), i.e. its complete utilization in oxidized sandstone strata as established by geologists.

The reasons of minerals micro dimensionality in uranous ores seems logical to consider as a consequence of mineral-forming environment mobility. In general case of classical (roll-front type) sandstone Deposits, lifetime of conditions for U deposition in a specific volume is of short duration (on a geological scale) due to motion of redox front. As a result, time interval for U-minerals formation (black ores) is limited; it depends on duration of redox front displacement along ore-bearing layer. The micron sizes of uranous minerals (crystals of coffinite and ningyoite in black ores) are considered as a consequence of mobility of reductive medium (redox barrier) that is moving along permeable ore-hosting stratum (Doynikova 2016). Microorganisms decompose buried organic residues and thus create a sedimentation environment for U^{4+}. The abundance of organic residues in the host rocks provide the more duration of their bacterial utilization, slowing the progress of redox front during water filtration. Therefore, the period of U-minerals deposition (and, accordingly, their size) is limited by the time of movement of reductive zone along the sedimentary layers. Progress promotion of redox front will depend on the amount of nutrients aerobically "prepared" (terminology Vinichenko 2004). Obviously, this is what determines the time of existence of uranium precipitation conditions in the reductive zone of redox front.

In the general case of classical roll-front deposit, the lifetime of the conditions for U deposition in a specific volume is of short duration (on a geological scale) due to the motion of redox front. As a result, the time interval for the U-minerals formation (black ores) is limited; it depends on the duration of redox front displacement along the ore-bearing layer. The micron sizes of U^{4+} minerals (crystals of coffinite and ningyoite in black ores) are considered as a consequence of the mobility of mineral-forming environment (redox barrier) that is moving along permeable ore-hosting stratum (Doynikova 2016). The abundance of organic detritus in the host rocks provide the more duration of their bacterial utilization, slowing the progress of roll front during water filtration.

6.2 Phosphorus Source for Ningyoite Formation

When clarifying the genesis of U^{4+}-phosphate mineralization, the issue of the phosphorus source is primarily considered. For a long time, the main unresolved issue was the source of phosphorus for the formation of ningyoite mineralization. Mineralogical observations in the review work (Boyle et al. 1981), as well as our data of numerous ningyoite samples EM studies, testify to non-essential role of clastic apatite as a source of phosphate in solutions, because there are no traces of dissolution of its' clastic grains, found in close association with ningyoite. Data of modern environmental mineralogy and geomicrobiology made it possible to understand this question and reveal the biogenic nature of ningyoite. Overview of the processes governing (manipulate) the interaction of U and P in contaminated soils looks in (Seaman et al. 2015).

Phosphorus, as one of the most important biogen elements, is a part of all living tissues, proteins and other organic compounds, participates in all types of metabolism (average P content in the living substance n-10^{-2}). Many processes of phosphorus concentration in the earth's crust are associated with biogenic migration; organisms absorb phosphorus from soils and aqueous solutions. According to A.I. Perelman, water migration of P is limited: "Only a small part of mobile P enters natural waters because it easily leaves them, being a part of insoluble minerals or being captured by organisms" (Perelman and Kasimov 1999). The established proportions of different forms of P coming from land to ocean (during year): 83% in composition of precipitation, 17% in form of soluble forms. This can probably be correlated with P accumulation in alluvial sediments.

Phosphorus enters sedimentary rocks (and soil) mainly with plant and animal residues in form of organic compounds, where P is present in the form of radicals of phosphoric acid, and in a changeless form, mainly as PO_4 complex, it is actively involved in various biochemical transformations (Makarov 2009). Phosphorus is a part of high-molecular humic acids (0.2% P in humic acid, 2–3 times more in organomineral complexes). The phosphate complex is extricated from organic and organomineral compounds by phosphorus bacteria, which convert organic phosphorus into inorganic compounds. Many microorganisms are able to transfer insoluble phosphate compounds into a soluble state (reverse process).

Assumptions of geochemists about appearance of phosphorus in ore strata as a result of microorganisms' activity (Kajitani 1970) or as a result of groundwater infiltration through organic precipitations (Boyle et al. 1981) have been convincingly confirmed in modern experiments by microbiologists. The processes of phosphate complex transformation in aquatic environment containing organic matter (i.e. in sedimentary deposits) are detailed in a number of works devoted to immobilization of radionuclides and bioremediation. Experimental studies (Lovley et al. 1991; Sivaswamy et al. 2011) show that bacterial fixation of uranium is usually in the form of phosphates. At the same time, phosphate ligands for formation of hard-to-soluble mineral forms U^{6+} and U^{4+} are formed by releasing phosphate complex from organic cells of microorganisms, which was observed during numerous experiments with different types of bacteria. Uranium has been shown reduced by enzymatic actions of microorganisms in these processes (Suzuki et al. 2005a, b).

Precipitation of oxidized form uranium (U^{6+}) together with biogenic phosphate ligands leads to accumulation of uranium complexes inside the cells or on their surface, or leads to formation of solid mineral deposits (such as autunite). Mineral forms of reduced uranium (U^{4+}) are noted both in abiotic precipitates (Behrends and Cappellen 2005; Bernier-Latmani et al. 2010) so in biotic cells. Under aerobic growth conditions, some facultative microorganisms may accumulate intracellular phosphorus (as polyphosphate granules) and under subsequent anaerobic conditions, these granules are hydrolyzed, releasing inorganic phosphorus (PO_4^{3-}) from the cells (Sivaswamy et al. 2011).

As shown in article (Suzuki et al. 2005a, b) microbiological experiments on example of uranium demonstrate that the source of phosphorus (P) in water solutions is organic cells; microbial reduction of U^{6+} can occur either directly by enzymatic action in presence of electron donor or indirectly by electron transmission. As a consequence of such transfer, resulting from humic acid, protons (H^{+}) appear in solutions. Here it is appropriate to recall specified above active role of H^{+} in ningyoite structure forming. It was shown at uranium example that biological deposition of radionuclides is always associated with microbial release of PO_4^{3-} from microorganism cells (Sivaswamy et al. 2011).

Microbiological experiments on the bacterial reduction of U (Behrends and Cappellen 2005; Cerrato et al. 2013; etc) demonstrate that the U^{4+}-mineral formation (uraninite, ningyoite) in natural waters saturated with organic matter is caused by the microorganisms' activity. At the same time, the source of phosphorus in natural solutions is the decomposition products of organic matter, mainly it is plant residues. Plant decomposition is ensuring supply of PO_4-ions to solution and is always accompanied by bacterial reduction of U^{6+}. So, microorganisms that disintegrate buried organic remains in the host strata provide for U mobility in form of uranyl-ions, organic complexes and colloids. The experiments on bacterial U-reduction have shown that during formation of ningyoite the source of phosphorus in natural solutions is decomposition products of organic matter. As a result, the amount of phosphorus (transformed mainly from plant remains) must be setting by content of plant detritus in ore-hosting stratum. The review of publications on uranium microbial immobilization (Sivaswamy et al. 2011) has shown that U fixation often occurs in form of phosphates (as U^{4+} and U^{6+}), and less often as uraninite.

Laboratory experiments on enzymatic and abiotic recovery of U^{6+} under iron-reducing conditions have shown that many sulfate and metal-reducing bacteria are capable of reduce uranium (Behrends and Cappellen 2005). Microbial reduction of uranium in processes of thermophilic (50–70 °C) iron reduction has shown (Slobodkin 2008) that microorganisms are able to use uranyl mineral (uramphite) as an electron acceptor for growth energy. Extracellular (enzymatic) reduction of U^{6+} to U^{4+}, as a result of bacterial growth, results in formation of ningyoite precipitate (Khijniak et al. 2005). This phosphate is considered to be a more common reduction product than uraninite, and in near-surface conditions it may be the dominant product of microbial reduction uranium (Bernier-Latmani et al. 2010).

It was be declaring that sulphate reduction, which is usually found in silt sediments or silted soils, are the dominant anaerobic terminal process in destruction of organic matter (Slobodkin 2008). Sulfate-reducing bacteria inhabit in bottom sediments of sea or occur in freshwater reservoirs rich in decomposing organisms; bacteriobenthos is the most abundant in silted soils. The leading role of microorganisms in the removal of radionuclides by microbial community of silts in water bodies has been traced in article (Khizhnyak 2013).

Lacustrine and wetland sediments and floodplain soils in humid landscapes, where phosphorus migrates and accumulates, represent a highly biologically active environment (Perelman and Kasimov 1999); here, similar to blacksoils, the amount of microbial phosphorus compounds can reach 80–90% of the total phosphorus content. Thus, the data of microbiologists explain close connection of ningyoite ores with lake-boggy sediments, which was identified by geologists and noted above.

Microbiologists' work on uranium immobilization by natural bacteria (in form of phosphate minerals) has convincingly shown that source of phosphate ions in aqueous solutions are plant cells, more precisely—products of bacterial recycling of various P-containing organic compounds in these cells. All above facts allow considering organic material of sediments (mainly plant detritus) as a source of phosphorus for ningyoite. Possible contribution of abiogenic dissolution of phosphates is insignificant.

In natural conditions, these conclusions are confirmed by ningyoite ores localize to strata most saturated with organic material. In hypergenesis region it may be alluvial paleochannel deposits or intensively worked out weathering crust, or paleostreambed alluvial sediments. Thus, geological observations and mineralogical data, as well as the results of experiments on bacterial reducing of uranium, lead to the conclusion: during ningyoite formation, *the source of phosphorus in natural solutions is the products of organic matter decay*. The amount of phosphorus transformed mainly from plant residues is limited by content of plant detritus in ore-bearing strata. By example of Vitim deposits, features of phosphate black U-ores localization and reasons for their close association with sulfides are revealed.

As shown above, only active participation of microbiota in decomposition of organic matter can explain both phosphorus release from plant residues into solution and so uranium mobility in the form of organic complexes and colloids. The source of phosphorus in ningyoite formation is mainly organic matter of plant residues (mortmass) in the sedimentary strata.

From data of geomicrobiology and microbiology it follows that vital activity of joint aerobic-anaerobic bacteria community within ore-hosting strata provides for U mobility and reducing environment during U black ores formation. Consequently, reason for various uranous mineral forms should be defined by features of microbiological processes.

6.3 Phosphate Uranous Ores Composition as Indicator of their Biogenic Origin

Biogenic aspect of black uranium ores formation is considered here on example of their phosphate component—ningyoite. In spite of this mineral wide study in domestic investigations, its origin and source of phosphorus remained insufficiently clarified. Publication of geomicrobiological researches results and scientific works on bioremediation connected with solution of ecological problems helped to solve the question about phosphorus source for phosphate black ores formation.

The formation of phosphate ningyoitic ores, when preserving the general regularities of U-ore formation in sedimentary strata, requires increased activity of phosphate ion, as noted in the early review of foreign works (Boyle et al. 1981).

The mechanism of this activity can probably be explained by the enzymatic action of microbial community that utilizes organic residues in the ore-forming medium. It is this provides for a lot of phosphate ions (their' increased activity) in the solution during the plant-remains decomposition in sedimentary strata. From the geological point of view, the main reason for appearance of phosphate uranous mineralization is significant predominance of plant residues (paleo-channel alluvium), i.e. products of their microbial transformation—humus organic substances—in the ore-bearing strata. Obviously, it is this dominance of the reworked flora products in the sedimentary strata that provides the "building material" (phosphate-ions) for ningyoite formation.

Similar processes also occur in the strongly reworking weathering crusts. A clear example is the Kosachinoe deposit, Northern Kazakhstan, where ningyoite ores are formed in the mature weathering crust what provides abundance of humus. As discussed above (Chap. 2), location of ores in powerful (up to 150 m) "pockets" of weathering indicates deep hypergenic reprocessing of rocks. Commercial ores are enriched with organic carbon (up to 15% S_{org}) in comparison with both primary ores and host rocks indicate a significant contribution of anaerobic bacteria in processing. To explain such significant contribution of anaerobic bacteria, a hypothesis of microbial community growth in anaerobic weathering zone (of ore forming) due to autotrophic bacteria that used primary carbonate mineralization of the ore-containing strata for their development was previously proposed (Doinikova 2007). Biogenic specific of genesis this ningyoitic ores points to that colloid environment have been created. In the terminal destruction of organic matter in silt sediments and silted soils, sulfate reduction is the dominant anaerobic process (Khijniak et al. 2005; Slobodkin 2008). This explains the close association of ningyoite and pyrite (noted earlier) in the strata with lacustrine-bog deposits, where ningyoite U-ores are localized.

A.I. Perelman established that the colloidal form of uranous migration is most typical for natural waters of hypergenesis region rich in high-molecular humus substances. Accumulation of colloids in biosphere is proportional to intensity and duration of processes of landscape development, allowing complete transition of all solids into colloidal and metacolloidal minerals in mature weathering crust (Perelman 1968; Perelman and Kasimov 1999*)*. Therefore, the formation of commercial ores from Kosachinoe deposit should be considered as a result of colloidal form of uranous (U^{+4}) migration, which is also known for other natural objects (Malkovsky 2011*;* Malkovskii et al. 2014; Malkovsky et al. 2015; Wang et al. 2014; Priyadarshini et al. 2014). Obviously, the formation of ningyoite ores in weathering crusts, as in Bulgaria and Northern Kazakhstan, allows us to speak about the *decisive role of colloids in* U^{+4}*-phosphate mineral formation.*

In sandstone type uranium deposits (of soil- or plast-infiltration) equally with ningyoite there is coffinite and amorphous pitchblende in ores composition (may be in subordinate amount). Such is commercial ore of Grachevskoye ore field (Kosachinoe); ores of stratiform deposits of the Czech Bohemian massif; ore bodies' composition in tufogene-sedimentary strata of Bulgaria (Navysen, Maritsa). When uraninite predominant of over ningyoite in ore composition (as in Sugraly,

Kyzyl-Kum) we have not found any carbonaceous organic remains; probably, their content was minimal.

Geological localization of ningyoite deposits corresponds to known regularity of increasing role of colloidal migration in humid landscapes in comparison with arid ones. The abundance of organic material and microorganisms in lake-boggy sediments, located among other sedimentary layers, provides formation of colloidal solutions in aquatic environment of ore-bearing strata. That is contributing migration of uranium and migration of phosphorus (as discuss above). From review publication (Doinikova 2007) it follows that the presence of lacustrine-bog sediments in hosting formation is characteristic for black phosphate ores. In the terminal destruction of organic matter in silt sediments and silted soils, sulfate reduction is the dominant anaerobic process (Khijniak et al. 2005; Slobodkin 2008). This explains the close association of ningyoite and pyrite (noted earlier) in the strata with lacustrine-bog deposits, where ningyoite U-ores are localized.

Japanese geochemists (Iwatsuki et al. 2003) have studied modern redox conditions at Tono uranium deposit, which is located in lower part of sedimentary Miocene rocks series, which unconformably overlapping granites. This deposit's ores are mainly composed of ningyoite. Buffer redox capacity of water-microbial system into and around this deposit was studied. Abundance and viability of microbes in these waters, as well as the isotopic ratios of sulphate sulphur, gave rise to conclusion that microbial sulfate-reduction involves organic matter in ore formation process. Together with subsequent pyrite deposition, these reactions are dominant at the depth of uranium-ore bodies.

6.4 The Reasons for Different Uranous Ores Composition

Our long-term experience of studying uranium black ores of various sandstone deposits has proved their polymineral character (Belova and Doynikova 2003; Doynikova 2003, 2012; Doinikova 2007; Velichkin et al. 2003, 2005; Doynikova et al. 2002, 2003, 2012, 2016; Doinikova et al. 2009). Ningyoite, uraninite (nasturane) and coffinite may occur in composition of black uranium ores in different proportions. The convincing identification of biogenic nature of phosphorus in ningyoite allowed us to details mineralogical processes occurring into medium redox barrier; discussing features of black uranium ores formation and causes of their different composition became possible.

Speaking about the biogenic aspect of black uranium ores formation, the answer to question about the reasons for diversity mineral form U^{4+}, obviously, should be sought in the peculiarities of microbiological processes, more precisely—in the peculiarities of chemical elements migration (when deposits with different ore composition are forming). According to A.I. Perelman (1968, Perelman and Kasimov 1999), climate is an external factor of supergene processes that's form (among other things) sandstone-hosted uranium deposits. Therefore, as a main cause of different composition of black uranium ore, we consider the difference in climatic conditions of sedimentation and the subsequent formation of U-ore strata formation.

By A.I.Perelman theories on the geochemical barriers and mobility of elements, the increased coming of phosphate ions in solution is characteristic of the humid climate. Moreover, at the same time, climate is an external factor in hypergenic processes that's form (among other things) sandstone-hosted uranium deposits. From geomicrobiology and microbiology data it follows that vital activity of joint aerobic-anaerobic bacteria community within ore-hosting strata provides for U mobility and the reducing environment during U black ores formation. Consequently, the reason for various mineral forms of U^{4+} should be defined by the features of microbiological processes.

It is discovered that the main reason for appearance phosphate uranous mineralization is significant predominance of humus organic substances in host-rocks. It is this dominance of the reworked flora products in sedimentary rocks ensures saturation of the environment by phosphate complex. Lacustrine-marsh deposits, typical for ningyoite ores, indicate a humid climate that leads to accumulation of excess stocks of mortmass (in comparison with arid climate) and an increased flowing of phosphate ions into solution during its decomposition. Similar processes take place in highly reworked weathering crusts, as on Kosachinoye, Northern Kazakhstan, with ningyoite (secondary) ores (Doinikova 2007). While such sediments were formed, *humid climate* provides for accumulation of phytomass and considerable reserves of mortmass (Perelman and Kasimov 1999). This circumstance provides for the increased level of phosphate ions (its increased activity) in the solution during decomposition of plant remains in sedimentary strata.

When there are few organic matter, as for example at sandstone-hosted (tabular type) Beverley uranium deposit in South Australia, the bulk of the mineralization consists of coffinite and/or uraninite (Wülser et al. 2011). The fine-grained uranium mineralization here was deposited mainly in organic matter-poor Miocene lacustrine sands. The prevailing conditions during the mineralization (pH 6.3 to 8.4, at 25 °C) provide essentially bacterial reduction. Some P and Ca in coffinite was tractate by the authors as the presence of admixtures in black uraniferous nodules. But most likely the presence here P-containing coffinite or phosphosilicate U^{4+} (coffinite-like) as in Dalmatovo (Doinikova et al. 2009); where really P and Ca are present in composition and ratio (U + Ca)/(Si + P) is close to 1 too, as for Beverly coffinite. The presence of phosphorous in U- ore mineralization Beverly is caused by increased activity of phosphate-ions due to deposition of Beverly ore sandstones in lacustrine shore environment enriched microbiota. But the amount of phosphate ions was probably insufficient to form ningyoite.

Let us compare the typical *sandstone-hosted* deposits with different U-ores composition. The *basal channel* KhOF deposits with phosphate-black U-ores are formed in paleo-stream alluvial deposits in a temperate and humid climate (Kochkin et al. 2017). The known large *roll-front* deposits of Near-Tien-Shan mega-province (Chu-Sarysu type) with uraninite-coffinite black ores are confined to the areas of arid climate, and are formed in sediments of the coastal sea basin, «continental to marginal marine basin» (Uranium 2016; World Distribution of Uranium Deposits

(UDEPO) 2016 2018). Likely, the climate-related abundant plant debris in the Vitim strata provided the predominant formation of U-phosphate black mineralization here, in contrast to the black U-ores of Central Asia.

While comparing various same-type basal channels (paleovalley) U-deposits, the reasons of difference in uranous mineralization are assumed to be in external ("nearly climatic") conditions. The examples are: the phosphate composition of KhOF uranium ores and nasturane-coffinite composition of Trans-Ural deposits (Dalmatovskoye, Khokhlovskoye etc). Probably, the external thermal effect of regional heating of the KhOF host-rocks by thick basalt covers became the reason of the difference in their ore composition. Such heating stimulated the vital activity of the bacterial community. This caused deeper terminal (final) processing of biomass and, as a result, increased activity of phosphate ions in the ore-forming water environment on Vitim deposits.

All said above show—role of microbiological factors in uranous ores formation is decisive. Modern up-date studies indicate the biogenic nature of the U^{+4}-mineralization in sandstone-type uranium deposits. A study of uraninite and coffinite from the Xinjiang sandstone roll deposit in China has shown that these minerals are precipitated by microorganisms (Min et al. 2005). A study of coffinites from Beverly uranium sandstone deposit in Australia concluded that bacterial reduction may be the primary ore-forming process (Wülser et al. 2011). Various geological publications recent years also indicate mineral-forming role of microorganisms in the formation of coffinite and uraninite in sandstone type deposits (in addition to the above—Cai et al. 2007; Zhao et al. 2018; etc). Current studies of biogenic uranium deposition in the processes of sulfate and iron reduction show that ore formation, accompanying bacterial activity, can continue until now (Iwatsuki et al. 2003; Behrends and Cappellen 2005; Slobodkin 2008). On the base of microbiological data together with our geological conclusions on biogenic nature of uranous phosphate, ningyoite, the decisive participation of biogen factor in black uranium ores formation can be considered as doubtless.

6.5 Geochemical Law Named after V.I. Vernadsky

The main scientific result of the author researches became a demonstration of action V.I. Vernadsky law in nature, precisely—in hypergenesis region. This law, one of the basic laws of Geochemistry, takes into consideration the General effect of the living matter activity during entire geological Earth history. All provisions/postulates of this law, formulated by famous Russian geochemist and soil scientist A.I. Perelman, about *direct participation of living substance* as *in geochemical activity* and so in *creation of medium* for chemical elements migration are appearing in formation of ningyoite. From the consideration of geologic-mineralogy and geomicrobiology data it follows that the vital activity of bacteria within ore-hosting strata provides for U mobility and creation of reducing environment during U black ores formation. As shown for ningyoite above, *only active participation of microbiota in decomposition of organic matter can explain both phosphorus release from plant residues into solution and so uranium mobility in form of organic complexes and colloids.* Review of literature on environmental Mineralogy have been shown that biogenic migration of uranium and phosphorous in the hypergene region (in Earth's crust) occurs in the environment whose reductive geochemical features are caused by living matter—microbial aerobe-anaerobe community. Together with the author's mineralogical research results, these data allow us to talk about decisive role of microorganisms in formation of uranous (U^{4+}) ores. This law is also applicable to other uranous minerals (coffinite and uraninite) that formed in similar hypergenesis conditions, where biogenic migration of tetravalent uranium is associated with organic derivatives of biosphere.

6.6 Conclusions

The peculiarities of phosphate-uranous ores formation demonstrate action one of the basic laws in Geochemistry hypergenesis—V.I. Vernadsky law. It's maybe the most considerable scientific result of conducted researches. A review of literature data on environmental mineralogy and the author's results in this aspect allow to talk about decisive role of microorganisms in uranous (U^{4+}) ores formation. As shown above, only active participation of microbiota in organic matter decomposition can explain both uranium mobility, in form of organic complexes and colloids, and phosphorus entrance into solution from plant residues. *The action of aerobic-anaerobic bacteria community leads to the creation of biogeochemical barriers (medium) that reduce uranium.*

Anaerobic reducing environment is caused by *summary* action of microbiological processes (microorganisms' community) in *integrated* redox barrier space, where aerobes and anaerobes coexist. It is assumed that promotion this medium (of joint biogeochemical redox barrier) along strata is occurring as nutrient base is consumed. The micron sizes of uranous minerals in black ores are considered as a consequence of mobility of mineral-forming environment (redox barrier) that is moving along permeable ore-hosting stratum together with roll-front.

The community of aerobic-anaerobic microorganisms should be evaluated as an important component of uranous ore formation along with other geochemical characteristics of environment. Based on microbial community unity in ecosystem, it is necessary to expand concept of "ore-localizing" barrier: its structure inherently includes not only reducing zone, but also oxidizing one.

In sandstone hosted uranium deposits, organic residues contained in host sedimentary strata always provide a nutrient medium for microorganisms. Essentially, *the very nature of sedimentary accumulation provides the presence of reductive agents that can create uranium precipitation barriers in the filtration of oxygen-bearing uranium solutions.*

References

Behrends T, Cappellen P (2005) Competition between enzymatic and abiotic reduction of uranium (VI) under iron reducing conditions. Chem Geol 220(3–4):315–327

Belova LN, Doynikova OA (2003) Formation conditions of uranium minerals in the oxidation zone of uranium deposits. Geol Ore Deposits 45(2):130–132. [In Russian]

Bernier-Latmani R, Veeramani H, Dalla Vecchia E et al (2010) Non-uraninite products of microbial U (VI) reduction. Environ Sci Technol 44:9456–9462

Boyle DR, Littlejohn AL, Roberts AC et al (1981) Ningyoite in uranium deposits of south Central British-Columbia, first north American occurrence. Canad Miner 19(4):325–331

Cai C, Li H, Qin M, Luo X, Wang F, Ou G (2007) Biogenic and petroleum-related ore-forming processes in Dongsheng uranium deposit, NW China. Ore Geol Rev 32:262–274

Calas G, McMillan PF, Bernier-Latmani R (2015) Environmental mineralogy: new challenges, new materials. Elements 11:247–252

Cerrato JM, Ashner MN, Alessi DS, Lezama-Pacheco JS, Bernier-Latmani R, Bargar J, Giammar DE (2013) Relative reactivity of biogenic and chemogenic uraninite and biogenic non-crystalline U(IV). Environ Sci Technol 47:9756–9763

Doinikova OA (2007) Uranium deposits with a new phosphate type of blacks. Geol Ore Deposit 49 (1):80–86. https://doi.org/10.1134/S1075701507010047

Doinikova OA, Solodov IN, Chertok MB (2009) Mineral composition of uranium ore at the Dalmatovo deposit, Russia. Geol Ore Deposit 51(6):505–514

Doynikova OA (2003) Genetic crystal chemistry of mineral components in uranium blacks. Geochem Int 41(12):1214–1220

Doynikova OA (2012) Mineralogy of uranium in the reduction zone of supergenesis (as revealed by electron microscopy). Publ H FIZMATLIT, Moscow, p 216. [In Russian]

Doynikova OA (2016) Phosphate composition of uranium sooty (blacks) as indicator of their biogenic nature. Proc higher educational establishments. Geol & Exploration 5:17–29. [in Russian with English abstract]. https://doi.org/10.32454/0016-7762-2016-5-18-25

Doynikova OA (2017) The biogenic aspect of the uranium blacks formation (manifestation of the law named after V.I. Vernadsky). In: Kasimov N, Gennadiev A (eds) Landscape geochemistry/ 100th anniversary of AI Perelman. Publ H APR, Moscow, pp 524–543. [In Russian]

Doynikova OA, Belova LN, Gorshkov AI Sivtsov AV (2002) The polymineral nature of uranium blacks. Uranium deposits: from their genesis to their environmental aspects (Proc. Intern. Workshop. URANIUM 2002) Czech Geol. Survey, Prague, 45–48

Doynikova OA, Belova LN, Gorshkov AI, Sivtsov AV (2003) Uranium blacks: questions of genesis and mineral composition. Geol Ore Deposit 45(6):452–465. [In Russian]

Erlich HL, Newman DK (2008) Geomicrobiology, 5th edn. CRC Press, Boca Raton, p 628

Hydrogenic uranium deposits (1980) Ed. Perelman AI. Atomizdat, Moscow: 270 [in Russian]

Iwatsuki T, Murakami Y, Naganuma T, Hama K (2003) Redox buffer capacity of sedimentary rocks around uranium deposit (study of water-mineral-microbe system at the Tono uranium deposit, Japan). Chikyukagaku (Geochemistry) 37:71–82

Kajitani K (1970) A geochemical study on the genesis of ningyoite, the special calcium uranous phosphate mineral. Econ Geol 65:470–480

Kalmykov SN (2008) The role of colloidal particles in the migration of actinides with groundwater. Diss.... Dr. chem. Sciences: 02.00.14. Moscow, 308 [in Russian]

Khijniak TV, Slobodkin AI, Coker V et al (2005) Reduction of uranium(VI) phosphate during growth of the thermophilic bacterium Thermoterra-bacterium ferri reducens. Appl Environ Microb 71(10):6423–6426

Khizhnyak TV (2013) Bacterial transformation and immobilization of heavy metals and radionuclides. Dis.... doctor of biological Sciences: 03.02.03. Moscow. 212. [in Russian]

Kislyakov YM, Shchetochkin VN (2000) Hydrogenic ore formation, vol 608. Geoinformark, Moscow. [in Russian]

Kochkin BT, Solodov IN, Ganina NI et al (2017) Geochemical features of the ore-bearing medium in uranium deposits in the Khiagda ore field. Geol Ore Deposits 59:341–353. https://doi.org/10.1134/S1075701517050026

Lezama-Pacheco JS, Cerrato JM, Veeramani H et al (2015) Long-term in situ oxidation of biogenic Uraninite in an alluvial aquifer: impact of dissolved oxygen and calcium. Environ Sci Technol 49:7340–7347. https://doi.org/10.1021/acs.est.5b00949

Lisitsin AK (1975) Hydrogeochemistry of ore formation. Nedra, Moscow, p 247. [in Russian]

Lovley DR, Phillips EJP, Gorby YA, Landa ER (1991) Microbial reduction of uranium. Nature 350:413–416

Makarov MI (2009) Phosphor of organic matter of soils. Moscow, GEOS, p 370

Malkovskii VI, Petrov VA, Dikov YP et al (2014) An analysis of the role of colloids in uranium transport in groundwater within the U-Mo deposits of the Streltsovsky ore field (eastern Transbaikalia region). Dokl Earth Sci 454:29–31. https://doi.org/10.1134/S1028334X14010024

Malkovsky V (2011) In: Kalmykov SN, Denecke MA (eds) Theoretical analysis of colloid-facilitated transport of radionuclides by groundwater. Actinide nanoparticles research. Springer, Berlin, Heidelberg, pp 195–243

Malkovsky VI, Petrov VA, Dikov YP et al (2015) Colloid-facilitated transport of uranium by groundwater at the U–Mo ore field in eastern Transbaikalia. Environ Earth Sci 73:6145–6152. https://doi.org/10.1007/s12665-014-3840-2

Marchik TP, Efremov A (2006) L. Soil science with the basics of crop production. Grodno State University, Grodno, p 249. [in Russian]

Markov A (2015) Birth of complexity/evolutionary biology today: unexpected discoveries and new questions. ACT: CORPUS, Moscow, p 527. [in Russian]

Min M, Xu H, Chen J, Fayek M (2005) Evidence of uranium biomineralization in sandstone-hosted roll-front uranium deposits, northwestern China. Ore Geol Rev 26:198–206

Perelman AI (1968) Geochemistry of epigenetic processes (Hypergenesis zone). Publ H: Nedra, Moscow, p 331. [in Russian]

Perelman AI, Kasimov NS (1999) Landscape geochemistry. Astrea 2000, Moscow. [in Russian]

Priyadarshini N, Sampath M, Kumar S et al (2014) Probing uranium (IV) hydrolyzed colloids and polymers by light scattering, Journal of nuclear chemistry. 2014:232967. https://doi.org/10.1155/2014/232967

Rare metal uranium ore formation in sedimentary rocks (1995) Collection of scientific papers. Science, Moscow, p 256. [in Russian]

Seaman JC, Buettner SW, Hyun-shik C (2015) Phosphate–uranium interactions in soils. In: Selim HM (ed) Phosphate in soil. CRC Press, pp 59–97

Singer DM, Farges F, Brown GE Jr (2009) Biogenic nanoparticulate UO_2: synthesis, characterization, and factors affecting surface reactivity. Geochim Cosmochim Acta 73(12):3593–3611. https://doi.org/10.1016/j.gca.2009.03.031

Sivaswamy V, Boyanov MI, Peyton BM, Viamajala S, Gerlach R, Apel WA, Sani RK, Dohnalkova A, Kemner KM, Borch T (2011) Multiple mechanisms of uranium immobilization by Cellulomonas sp. strain ES6. Biotechnol Bioeng 108(2):264–276

Slobodkin AI (2008) Thermophilic iron-reducing prokaryotes. Dis. Dr. Biol. Sciences: 03.00.07 (microbiology). Inst Microbiology RAS, Moscow, p 336

Southam G, Sanders JA (2005) The geomicrobiology of ore deposits. Econ Geol 100:1067–1084

Strakhov NM (1957) Stages of sedimentary rocks formation and tasks of their study. In: methods of studying sedimentary rocks, vol 1. Gosgeoltehizdat, Moscow, pp 7–28. [in Russian]

Suzuki Y, Kelly SD, Kemner KM, Banfield JF (2005a) Direct microbial reduction and subsequent preservation of uranium in natural near-surface sediment. Appl Environ Microbiol 71 (4):1790–1797

Suzuki Y, Kelly SD, Kemner KM, Banfield JF (2005b) Direct microbial reduction and subsequent preservation of uranium in natural near-surface sediment. Appl Environ Microb 71 (4):1790–1797

Uranium (2016) Resources, production and demand. OÉCD (2016) NEA/IAEA/:7301

Velichkin VI, Kushnerenko VK, Tarasov NN et al (2005) Geology and formation conditions of the Karku unconformity-type deposit in the deposit Karku in northern Ladoga region (Russia). Geol Ore Deposit 47(2):87–112. [In Russian]

Velichkin VI, Tarasov NN, Andreeva OV et al (2003) Geology and formation conditions of Karkhu deposit–the first uranium "unconformity" deposit in Russia. Proc. AN "URANIUM GEOCHEMISTRY 2003" Nancy, France: 371–374

Vinichenko PV (2004) Theory of biogenic ore formation exemplified by uranium deposits. Sosnovgeologiya, Irkutsk. [In Russian]

Wang Y, Frutschi M, Bagnoud A, Suvorova E, Mcgivney E, Chesaux L et al (2014) Geochemical controls on the formation of U(IV)-bearing colloids in a mining-impacted natural wetland. Goldschmidt Abstracts:2647

Wang Y, Frutschi M, Suvorova E, Phrommavanh V, Descostes M, Osman AA, Geipel G, Bernier-Latmani R (2013) Mobile uranium (IV)-bearing colloids in a mining-impacted wetland. Nat Commun 4. https://doi.org/10.1038/ncomms3942/

World Distribution of Uranium Deposits (UDEPO) 2016 (2018) IAEA-TECDOC-1843, IAEA, Vienna

Wülser P-A, Brugger J, Foden J, Pfeifer H-R (2011) The sandstone-hosted Beverley uranium deposit, Lake Frome Basin, South Australia: mineralogy, geochemistry, and a time-constrained model for its genesis. Econ Geol 106:835–867

Zammit CM, Shuster JP, Gagen EJ, Southam G (2015) The Geomicrobiology of supergene metal deposits. Elements 11:337–342

Zavarzin GA, Kolotilova NN (2001) Introduction to natural-environmental microbiology. Publ H University, Moscow

Zhao L, Cai C, Jin R, Li J, Li H, Wei J, Guo H, Zhang B (2018) Mineralogical and geochemical evidence for biogenic and petroleum-related uranium mineralization in the Qianjiadian deposit, NE China. Ore Geol Rev 101:273–292. https://doi.org/10.1016/j.oregeorev.2018.07.025

Conclusion

The monograph from different sides has been discussing the previously little-studied black uranium-ore mineralization, which is manifested in the form of a loose powder in almost all types of uranium deposits. So-called uranium blacks (sooty) are composing the ores of the Sandstone-type uranium deposits most demanded by industry that predominate in the world's uranium production. In recent years, more than half world uranium production was extracted from deposits of this industrial type.

Uranium blacks are formed in the weathering crusts of sulfide-containing uranium ores in the lower horizons of activity hypergenic redox processes (secondary enrichment zones), as well as in hydrogene uranium deposits of sedimentary cover. The ores of hydrogene deposits, in fact, represent an extensive mineralization of the reducing zone hypergenesis. The reducing factor in the development of the hypergenic process is mainly microbiota of sedimentary strata; the vital activity of this bacterial community is provided by the presence of organic matter in the ore-bearing strata. This is more obviously appearing in the formation of black mineralization in various infiltrations deposits; here uranous ores are everywhere composed of uranium blacks.

Conditions and regularities of hypergenic uranium mineral-formation are considered. The place of reductive (black) mineralization in the systematics of exogenous *ore* concentration of uranium, created by Russian mineralogist L.N. Belova, is shown. For the first time, stages of the processes of uranium mineral-formation in hypergenesis region have been schematized; the diagram shows all types of ore concentration uranium, united by common (redox) laws of mineral formation.

Geochemical aspects in this work are discussed based on researches of Russian geochemical and soil scientist A.I. Perelman and confirm him conclusions about the indissoluble unity of redox zonality in the entire zone of hypergenesis. It is advisable to consider the oxidation zones of hydrogenic deposits as an integral part of uranium ore-localizing biogeochemical barrier.

Schematic examples of uranium deposits of various genetic types, where blacks' mineralization is developed in commercial volumes, are given. These fields are:

O. A. Doynikova, *Uranous Mineralogy of Hypergene Reduction Region*,
Springer Mineralogy, https://doi.org/10.1007/978-3-030-67183-9

Taboshar, Central Asia; Chu-Sarysu type, Near Tien Shan; Ningyo-Toge, Japan; Kosachinoe, Northern Kazakhstan; Khiagda, Russia. The paleo-channel uranium deposits of Khiagda (Vitim) ore region are described in detail. A new for Russia, type of uranium-ore mineralization has been discovered here - phosphate black (ningyoite) one; it is of practical importance. The biogenic source of phosphorus for uranium ores of paleo channel deposits suggests the biogenic nature of redox barrier that precipitating uranium in sandstone hydrogenic deposits of hypergene region.

Mineralogical analysis of black uranium ores matter was first carried out at electron microscopy level by local crystallochemical methods. The methodological chapter shows how the use complex methods of analytical transmission electron microscopy (ATEM), based on electron microdiffraction SAED, allows us to study fine-dispersed, poorly crystallized and metamict uranium minerals at a qualitatively new level. The examples show how EM serves as the most effective tool for study of loose uranium ores, adding information into crystallochemical mineralogy of tetravalent uranium. Thanks to these ATEM crystallochemical data, a significant amount of fundamentally new information on uranous Mineralogy has been obtained. A new mineral family of uranous phosphates with a new mineral group of lermontovite was discovered, and the crystallochemical systematics of its mineral species was carried out. The systematics of Rhabdophane's group phosphates has been clarified (the list of its constituent minerals shortened). A number of new mineral varieties have been identified (Fe-ningyoite, P-coffinite, etc.). Complex of analytical EM methods (transmission and scanning) proved the polymineralty of uranium blacks; at least three uranous phases (uraninite, coffinite and ningyoite) can be including in its composition. The mineralogical section contains unique reference material, and aggregates all known data on the finds of previously unknown (little known) uranous phosphate ningyoite.

Significant progress in uranous minerals study is provided by transition of mineralogical research from optical microscopy level to electron microscopy one. The use of analytical electron microscopy methods raises mineralogical research of uranium ores to a new level.